Introduction to Rocket Science and Space Exploration

The growing demand of space services for imaging, mobile communication, global positioning systems and disaster management, life extension of satellites by fueling, space station operations, deflecting incoming asteroids, and reducing debris from orbits, requires reusable rockets. The chapters in the book cover understanding of the universe, history of rockets, space missions, satellites, the principle of rocketry, its design and development, rocket technology, the solar system, the environment and protection of earth, and thoughts on Earth 2.0.

Features:

- Explores the link between the universe, space exploration, and rocketry.
- Discusses topics such as protection of the Earth from asteroids, debris, and global warming.
- Includes basic methodology to be adopted to design rockets for various applications.
- Covers use of multi-objective optimisation to realise a system and differences in design philosophies for satellite launch.
- Examines material on environmental protection of the Earth.

This book is aimed at senior undergraduates and professionals in aerospace engineering.

Introduction to Rocket Science and Space Exploration

A. Sivathanu Pillai

CRC Press
Taylor & Francis Group
Boca Raton London New York

CRC Press is an imprint of the
Taylor & Francis Group, an **informa** business

Designed cover image: Polar Satellite Launch Vehicle is the workhorse rocket of the Indian Space Research Organisation (ISRO), for injecting 1000 kg class earth observation and navigation satellites in near earth or polar orbits. More than 50 successful launches have taken place from Sriharikota Satish Dhawan Space Centre.

First edition published 2023
by CRC Press
6000 Broken Sound Parkway NW, Suite 300, Boca Raton, FL 33487-2742

and by CRC Press
4 Park Square, Milton Park, Abingdon, Oxon, OX14 4RN

CRC Press is an imprint of Taylor & Francis Group, LLC

ISBN: 978-1-032-34697-7 (hbk)
ISBN: 978-1-032-34698-4 (pbk)
ISBN: 978-1-003-32339-6 (ebk)

DOI: 10.1201/9781003323396

Typeset in Times
by MPS Limited, Dehradun

Dedication

*This book is dedicated to
my three great teachers,
Dr. Vikram Sarabhai,
Prof. Satish Dhawan,
and Dr. APJ Abdul Kalam,
and to all those space pioneers and technologists
of the world.*

Contents

About the Author...xv
Foreword..xvii
Preface..xix
Acknowledgements..xxi

Chapter 1   Introduction...1

Chapter 2   Origin of the Universe 3
 Verse from the Ancient India 3
 Hindu Puranas...5
 Great Indian Saints (Scientists)...6
 Emerging New Scientific Thoughts on the Universe..................8
  General Theory of Relativity 9
  Einstein's Equations 10
  Proving Einstein's Theory.. 12
  Proved Einstein's Theories... 12
  Gravitational Lensing 13
  Gravitational Waves 13
  Cosmology 13
  Big Bang Theory 14
  CERN – Finding the GOD Particle 14
  Universe after BIG BANG... 15
  The Redshift – Our Expanding Universe 16
  Fundamental Physical Forces.. 17
  Gravity 18
  Electromagnetic Force... 18
  The Weak Nuclear Force 19
  The Strong Nuclear Force 19
  Unifying Nature.. 19
  Braneworld.. 20
 Muons 20
  String Theory... 20
  Stellar Evolution – Lifecycle of a Star............................... 21
  Black Hole 22
  Dark Energy and Dark Matter 25
   What the Universe Is Made Of.................................... 25
  Dark Energy... 25
   Background to Dark Energy 26
  Burst of Supernova... 26
  The Dynamics of Universe – It Will Never Shrink................ 28

New Dimension 28
Dark Matter... 28
Multi-Universe 30
Milkyway and Solar System 31
Solar System and Planets.. 32
KEPLER'S LAWS – Orbits of Planets................................... 32
The Blue Dot – The Earth 34
Origin of Earth 35
Life on Earth... 37
Future of Earth 39
Exoplanets... 40
Data to Remember... 40

Chapter 3   Brief History of Rockets 41

Early Understanding 41
Chinese Fire Arrows.. 42
First Metal Rockets from India... 42
Birth of Modern Rocketry... 44
Pioneer of Cosmonautics: Equation...................................... 44
Early Rocket Designs 44
Tsiolkovsky, the Father of Astronautics... 45
Tsiolkovsky's Rocket Equation 45
Robert H Goddard (1882–1945) 47
Hermann Oberth (1894–1989) 49
Space Odyssey... 51
Orbiting Satellite... 51
Yuri Gagarin 53
Man on Moon 54
Space Shuttle 55
Space Stations 55
China's Tiangong Space Station.. 56
Satellites in Brief (Details in Separate Chapter)............................. 56
Earth Observation Satellites.. 56
Communication, Navigation.. 56
Satellite Launch Vehicles.. 57
ISRO 57
Launch Vehicles of the World... 58
World Rockets at a Glance 61

Chapter 4   Rocket Principles... 63

What Is a Rocket? Basic Rocket Science 63
Principle 63
Newton's Laws of Motion 65
Newton's First Law – "The Law of Inertia"........................... 65
Newton's Second Law – "F = ma" 66

Newton's Third Law – "The Law of Action and Reaction"... 68
Forces on a Rocket during Initial Flight 68
    Centre of Gravity and Centre of Pressure 69
The Rocket Equation 70
    Thrust of Rocket 70
    Acceleration 72
    Total Impulse 73
    Specific Impulse 73
    Incremental Velocity – The Ideal Rocket Equation
    (ref. NASA) 74
        Valid Assumptions for an Ideal Rocket 77
Staging/Multistage Rockets 78

Chapter 5   Rocket Systems Development 81

Rocket Subsystems and Components 81
Propulsion 83
    Solid Propellant Rocket 83
    Large Solid Propellant Booster 83
    Burn Rate 85
    Liquid Propellant Rockets 87
        Pressure-Fed System 88
    Types of Liquid Propulsion 89
Cryogenic Rocket Engines (LOX +LH2) 90
    Semi-cryogenic Rocket Engines 92
    ISRO Semi-cryogenic Engine (SCE-200) 93
Liquid Ramjet Propulsion 96
    Supersonic-Combustion Ramjet (Scramjet) 98
    Scramjet Research 99
Mach Number vs. Specific Impulse 100
    Re-usability of Booster 100
Structure 101
Mass Fraction 101
Materials 103
Upper Stages 104
Casing for Liquid Propellant Stages 105
    Aluminam-Lithium Casing for Falcon 9 (2010) 105
Control System 105
    Types of Control 106
Different Types of Controls 107
    Control for PSLV Type LV 109
Satellite Attitude Control 109
    Flex Nozzle 109
Navigation and Guidance 110
Function 111
Inertial Guidance System 111

SINS (Stabilised Platform INS)...111
Guidance Sensors Accuracy and Applications...............................114

Chapter 6 Rocket Design Methodology, Test and Evaluation,
World Launch Sites..115

Launch Vehicle Design Methodology 115
SLV-3...116
    Launch Vehicle Design – A Simplified Form........................116
    Rocket Systems...117
    Check-Out...118
    Control System 118
    Guidance 118
    Trajectory Design 118
    Sequence of Tasks Before Flight..118
    Development of Design and Operations Software
    for SLV-3..119
Trajectory Design of ISRO's Mars Mission 120
    Geo Centric Phase..120
    Helio-Centric Phase..120
    Martian Phase...121
Human Mission to Mars (Inter Planetary Travel).........................122
    Wind Tunnel Testing and CFD 123
    The Final Choice is Space Launch System (SLS)
    for Moon, Mars 123
Van Allen Radiation Belts 124
World Launch Sites..124

Chapter 7 Satellites, Orbits and Missions 129

Satellites...129
SPUTNIK-1 and the Dawn of the Space Age 129
Importance of Satellites...130
Orbit..130
    Orbital Velocity of Satellite...131
    Minimum Orbit – First Cosmic Velocity 131
Satellite Orbits 132
    Low Earth Orbit (LEO)..132
    Medium Earth Orbit (MEO) 133
    Polar Orbit 133
    Geo Stationary Orbit (Arthur C. Clarke Orbit)....................133
    Velocity Requirement for Orbits and Escape from Earth......136
Satellites in Orbit...137
Mission Planning 137
Satellite Applications...138
Earth Observation Satellites 139
Navigation..140

Indian Regional Navigation Satellite Systems
(IRNSS) – NavIC 140
Small Satellites 141
Space Environment Hazards 142
Private Space Operators 142
Space-X 142
Starlink 142
Star Ship 143
Blue Origin 143
Other Types of Satellites 143
Human Space Missions 146

Chapter 8   Advances in Space Technologies 149
Can We Travel Faster than Light? 149
Electric Propulsion 150
The Need 150
Types of Electric Propulsion 151
Electrothermal Engine 151
Resitojet 152
Electromagnetic/Plasma Engine 153
MHD Engine 153
Electrostatic Thrusters 153
Ion Thruster 154
Hall Effect Thruster 155
HET Principle of Operation 156
Propellants 157
Xenon 157
Krypton 157
Cylindrical Hall Thrusters 158
External Discharge Hall Thruster 158
Nanotechnology in Electric Propulsion Thrusters 159
Magneto Plasma Rocket Propulsion (NASA) 160
Nuclear Propulsion Rockets for Space Missions 162
A NERVA Solid-core Design 162
Liquid Core 164
Gas Core – Closed Cycle and Open Cycle Figure 8.9 164
Bimodal Nuclear Thermal Rockets 165
Solar Electric Propulsion (SEP) 165
Laser Propulsion 167
Solar and Laser Powered Interstellar Sail 167
Hypersonic Transportation 168
Low-Cost Access to Space 170
Clean Energy Through Space 171
Artificial Intelligence and Robotic Spacecraft
Development 172

Autonomous Systems 172
NASA's Autonomous Science Experiment 173
Artificial Intelligence Flight Adviser 173
Stereo Vision for Collision Avoidance.................................. 173
Benefits of AI 174
Robotic Spacecraft Development... 174
Radioisotope Thermoelectric Generators.............................. 174
Establish Lunar Industry 174
Space Frontiers 176

Chapter 9   Environment: Protection of Earth and Sustainable
Development.. 177

Challenges for the Space Community 177
Threats from Outer Space 178
Protection of Earth from Risky Asteroids 179
Asteroids Collision Avoidance Strategies 179
Clearing the Space Debris... 180
New Business Venture 183
Space Systems Awareness and Comprehensive Space Security ... 183
Space Weather 184
Man-Contributed Threats on Land... 184
Population Increase.. 184
Global Warming 185
Consequences of Global Warming 187
Ozone Depletion 187
CO_2 Concentration 188
Temperature Increase 189
Temperature Profile 190
Positive Actions... 190
Geospatial Technologies for Sustainable Development
and Empowerment.. 191
Agriculture 191
Precision Farming.. 192
Basic Technologies Used in Precision Farming.................... 193
Information on Growth Stages.. 194
Field Zoning Based on the Productivity Level 194
Internet of Things 194
Geographic Information Systems.................................... 194
Fishing 195
Water and Land Resources 196
Bhuvan – A Unique Gateway to Indian Earth
Observation Data and Services... 196
Geo-Spatial Pyramid.. 197
Clean Energy Generation 197

Chapter 10 Exoplanets, Earth 2.0 199

         Exoplanets in Liveable Zones 199
            KEPLER-186f 199
            KEPLER-62f 200
            KEPLER-62e 201
            KEPLER-69c 202
            KEPLER-22b 202
            TRAPPIST-1 203
         Probable Earth 2.0 206
            TRAPPIST-1 Planets 206
            Proxima b 207
            TOI 700 d 208

Epilogue 209

Glossary 211

References 215

Index 223

About the Author

 After his graduation in engineering in 1969, **Dr. Sivathanu Pillai** worked in ISRO and DRDO during the last five decades, under aerospace leaders Dr. Vikram Sarabhai, Prof. Satish Dhawan, and Dr. APJ Abdul Kalam. He was also visiting professor in IISc and IITs. His active participation in the development of satellite launch vehicles and guided missiles made him specialise in multi-disciplinary technology projects, rocket technology, aerospace systems, institution build-up and advanced system engineering, and program management. Dr. Pillai's skills in networking multiple institutions for the development of critical missile technologies overcoming denial regimes (Missile Technology Control Regime) enabled realisation of strategic missiles. Dr. Pillai was the architect and founder CEO & MD of BrahMos Aerospace, which is a unique joint venture, the first of its kind between India and Russia to design, develop, produce, and market the most advanced supersonic cruise missile, BRAHMOS, which has been inducted in the Indian Armed Forces as a major strike weapon.

Fifteen academic institutions in India and abroad have recognised Prof. Pillai's scientific contributions and awarded him a Doctor of Science (Honoris Causa). He is the recipient of many prestigious national and international awards and fellowships, most importantly Padma Bhushan (2013) and Padma Shri (2002) by the Government of India, and the Order of Friendship (2013) by the Government of the Russian Federation. Prof. Pillai has written ten books, including "The Success Mantra of BRAHMOS," "40 years with Abdul Kalam – Untold Stories," "Revolution in Leadership," and "Nano Science and Nano Technology for Engineering." He co- authored with Dr. Abdul Kalam the books (a) "Envisioning an Empowered Nation" and (b) "Thoughts for Change – We Can Do It" (which defined 10 futuristic technologies).

Foreword

Planet Earth, moving in the vast universe, is also a place where life has evolved. The human race has been continuously endeavouring to improve its understanding of the universe. In this process, humans have developed tools to negotiate issues of survival. Initial periods of the survival struggle were spent in negotiating with other life forms. Even as the ability of humans developed to deal with various aspects of survivability, improving their living conditions and living standards, conflicts among human groups became dominant. These conflicts also resulted in intense efforts to develop technological capabilities to help them dominate over other groups. The ability to move quickly from one location to another became an important capability. Taming animals to ride on them and using them to pull carts gave a significant advantage to one group over the other. Developing vehicles to move on land, water, and air gained importance, and those who brought them to use gained dominance over other groups. These transportation modes not only enabled movement of people but also materials and products, enabling trade and commerce. However, each new capability also got utilised in establishing supremacy against those who were less capable. Thus, this process of developing newer and more capable transportation systems became a continuous endeavour of humankind. Using explosions to hurl weapons over long distances became one of the techniques adopted by people to dominate over opponents during war. Evolution of these explosive systems gave rise to capabilities of humans to put objects in outer space, orbiting around the Earth and through this bring in capabilities to use "space for Earth" and "space for space".

Today, humanity has benefited from the ability to use propulsion to carry objects to space and use these objects to perform communication, broadcasting, remote sensing, and navigation activities. The development has also enabled humans to go beyond Earth, orbit around the Earth, and land on the Moon.

While for decades the space activities were only handled by national agencies the world over, entry of private enterprises has changed the scope and pace of activities. Today space exploration, space adventure, space tourism, space exploitation, and use of space systems for asserting supremacy have all become a reality.

The author of this book, who has spent time and effort building propulsion-based systems for carrying objects to space that can be used for a variety of applications, has now concentrated his efforts on bringing to the reader a free-flowing narration on the many aspects of rocket science and space exploration. He has brought out the history of rockets wherein he traces how what transpired in the fight between the kingdom of Mysore and the British contributed to the development of rockets in the 18th century in Europe. The narration encompasses aspects of rocket principles, rocket systems development, rocket design methodology, and advances in space technologies. The author has brought out how rocket technology development has resulted in satellite missions covering different applications. Use of space system applications encompasses practically all aspects of human endeavours through the use of communication, broadcasting, meteorology, remote sensing, and navigation features.

While it is clear that technological progress keeps influencing the human thought processes and desires, the converse of this is also true. Konstantin Tsiolkovsky said in the late 19th century, "The Earth is the cradle of Mankind, but one cannot eternally live in cradle". Today, Elon Musk not only talks of making human beings a multiplanetary species but also working towards this objective. The author's narrative encompasses aspects of the beginning of life on Planet Earth in the first chapter and brings to us issues affecting the environment of Planet Earth, sustainability of development, and threats to life on Earth. The narration also brings to us efforts to look for Earth 2.0.

The author deserves to be complemented for collecting in one place such a rich compilation of information on rocket technology, the impact of space technology over life on planet Earth, and the search for Earth 2.0.

I am sure this book will be well received by the student community in particular and others in general.

आ श्री किरण कुमार,

(आ. सी. किरण कुमार)

(A. S. Kiran Kumar)
Vikram Sarabhai Professor (ISRO)
Former Secretary, DOS
Bangalore
22 December 2021

Preface

While in Russia, I had an opportunity to visit the residence of Konstantin Tsiolkovsky, in Kaluga, 140 km from Moscow. His residence is now a museum, where I sat in the century-old chair once used by him and looked through his hand-made telescope to the sky. It was inspiring to know that for the reason of his hearing impairment, he dropped out of school as a student but later became a schoolteacher. He gave mankind a vision to land on Mars and colonise other planets. He designed a rocket in the year 1886 and gave the rocket formula, long before man could fly through the air. The aura of his study room was nothing less than a temple to me. His quote *"Earth is the cradle of humanity, but one cannot remain in the cradle forever"* correlates with mankind's endeavour for the search of new Earth-like planet. NASA's Perseverance mission is one such attempt to study our neighbouring planet for its habitat suitability to mankind. That means human civilisation is soon on the verge of spreading its wings beyond Earth. While the vision of Tsiolkovsky is becoming reality, humans are taking adventurous steps to explore space, even beyond Mars.

While I was an engineering student, I participated in a science exhibition to display a unique processer, which attracted luminaries, Sir CV Raman and Dr. Vikram Sarabhai. The touch of Sarabhai, putting his right hand on my shoulders, electrified me to commit myself to join ISRO in 1969. My 40 years of long association with Dr. APJ Abdul Kalam and interaction with Prof. Satish Dhawan made me work in the field of rocket science and aerospace engineering, particularly in SLV-3, PSLV configuration at ISRO, and later in the integrated guided missiles development program at DRDO. They groomed me to work with rockets and later missiles, and that's how BrahMos came into being. My journey as a rocket scientist has posed various challenges and opened several avenues of possibilities.

Before joining ISRO, I had no knowledge of rockets. We had no exposure to the scientific advancements due to poor communication means. Today, space science has become a very important field, and basic knowledge is necessary for every school student. But even with better exposure to the world and connectivity, the current Indian education system does not promote creativity to young minds to think and innovate. School children receive little background on rocket science. Hence, this book is an attempt to enlighten, entertain, and inspire future generations on rocketry and understanding of the Universe.

When these children go for higher studies in colleges, they will become a new generation entrepreneurs in various niche areas of space opportunities. Space exploration, space tourism, colonisation of planets, emerging exotic technologies, and bio-needs of living in exo-planets will all demand a new breed of young innovators for many years to come.

The growing demand of space services – for imaging, mobile communication, global positioning systems, disaster management, life extension of satellites by fuelling them in orbit, space station operations, deflecting incoming asteroids, and reducing debris from earth orbits – all need low-cost reusable rockets and innovative solutions. So, basic knowledge this science must be imparted in schools.

ISRO envisioned by Vikram Sarabhai and carefully nurtured by Satish Dhawan has shown extra-ordinary results to make India one of the lead space-faring counties in the world. ISRO encourages university students to build nano satellites to be launched in PSLV piggy-back at no cost. This has highly motivated students, and more than ten such satellites have been put in orbit. Recently 1000 school children made 100 Femto satellites and launched them at high altitude by helium balloons and received good data through telemetry. As I was witnessing these developments, I could see the excited faces of the children. I found there are very few books on rockets, mostly written many decades before, and they are being made available in re-edited versions. But these books contain theories and formulae which will be of use for specialists who design rockets. Hence, the need arises to introduce rocket science to beginners, undergraduates, and school students. So, I covered only essential formulae and calculations. Moreover one needs to understand the use of rockets to orbiting spacecraft and telescopes for space exploration and to understand the Big Bang theory of the Universe.

The chapters contained herein deal with Understanding the Universe, Evolution of Rocketry, The Rocket Principle, Rocket Systems Development, Design Methodology, Satellites, Orbits and Missions, Advanced Rocket Technologies, Protection of Earth and Geospatial Technologies, and Thoughts on Earth 2.0.

I hope that this book shall appease the curiosity of young student learners, fledgling space enthusiasts, and engineering and science graduate students of different disciplines.

Acknowledgements

The author is thankful to NASA, ESA, ISRO, Ruscosmos, and other space organisations and web sites; Pinterest, Space.com, and Wikipedia; and the authors of many rocket science books given in reference, for valuable inputs to write the book. I relied most heavily on these inputs to write this book. I acknowledge ISRO and DRDO for giving me the opportunity to work with great rocket scientists and missile technologists, who excelled in their respective areas. Many valuable ideas came from them. NASA's rocket science courses designed for school children and educator guides have been a source of inspiration and guidance to write this book. I acknowledge with gratitude to NASA for those valuable inputs.

I express my gratitude to Dr. A.S. Kiran Kumar, former Chairman ISRO, with whom I worked at ISRO HQ recently, for giving a Foreword to this book.

I would like to acknowledge the objective reviews of the manuscript by Prof. G Jagadeesh of the Indian Institute of Science and Prof. Narahari K. Hunsur of the Department of Aerospace Engineering of MS Ramaiah University. Mr K Vijay Anand, DRDO, and Mr Hariharan, BrahMos, worked with me closely to edit and shape the book. I would like to express my gratitude profusely to both. I also thank Mrs. Surekha Kaul, BrahMos Knowledge Centre, and many of my friends for giving me valuable suggestions at different times while writing this book.

1 Introduction

Over 100 billion people lived on the Earth in the last 50,000 years. For many millennia, they looked at the skies, wondering at the stars and worshipping the visible Sun and the Moon. As early as 3000 BC, the Indian saints talked about the cosmic wonders, energy transformation, and even space travel. But during the last three centuries, scientific understanding of the mysteries of the cosmic arena and theories to substantiate the formation of the universe emerged. The advent of large telescopes and spacecraft launched through rockets enabled humans to explore new space frontiers that had mystified humankind. Rockets have become commonplace in our everyday use, and satellites are accepted as a necessity for life, generating excitement and enthusiasm. Their applications include television, movies, video games, weather reports, and books. Particularly, in the post COVID-19 world, life has been completely transformed to the digital world, and space applications have become more relevant for webinars, virtual classes for children, and conferences connecting different parts of the world.

I recall the days in July 1969 when I was glued to the radio (as there was no television at that time) listening with excitement to the running commentary of Neil Armstrong landing on the Moon. It was just after I sent my application to join ISRO, on completion of my engineering degree. I dreamt one day an Indian would land on the Moon. It is heartening to see now, spacecraft landing on the Moon and Mars, including the findings of Perseverance rover and astronauts spending months orbiting in the space station. These events are exciting to young and old people. Spacecraft launched by rockets go long distances to survey the planets, the Sun, the Moon and galaxies, as well as exoplanets for humans to settle later. So, we need to know more about the significance of rockets and emerging space technologies to explore the universe. That is why the book starts with a chapter on the universe, which is still unknown.

We must understand the evolution of rocketry and the subsequent developments that led to the fascinating space age of today. Rockets have become a means for humans to achieve great adventures to the unknown, since the days of Sputnik. A big rocket carrying a satellite has its mission of reaching its altitude using multiple stages and imparting the required velocity to inject the satellite precisely into orbit. Such a rocket has something of the order of one million parts. The reliable working and functioning of each of these parts must be ensured for the mission's success. So, rocket science is not a simple subject by any means; otherwise, the old joke about "it is not rocket science" would not be as funny as it is. But the basic principles of rockets can be easily understood by senior grade-school students of the modern age. With that view, this book attempts to discuss rockets and how they were discovered, developed, and used for different applications to improve the quality of life and to explore the universe.

DOI: 10.1201/9781003323396-1

Humans always want to know the future in advance and want to control their destiny. It is intriguing to know that an astrologer predicts our future based on the movement and positions of celestial bodies, with reference to our birth stars. In Hindu mythology, walking around the statues of eight planet Gods with the Sun God at the centre is believed to symbolise circling the real planets and Sun, and thus helps the person evade imminent bad happenings in life. The question is how the far-away stars can influence our lives on Earth, which is another planet. In temples, when I walked around the eight planet Gods and the Sun at the centre, I used to think that I am going around the solar system so easily. I have done this more than 1000 times, but in real life, I went around the true Sun 74 times. What is the science behind the influence of planets on Earth? When we study the Sun, planets, and the Moon, their effect of gravitation, magnetic forces, the fusion energy from the Sun, and vibrations of the strings have mutual influence on the Earth. I am convinced of the understanding of the accurate predictions of celestial events, based on the instantaneous positions of the planets, especially Jupiter and Saturn – the large planets and their influence on us. Indians practised astrological science thousands of years and predicted accurate happenings of events. As early as 5000 years before, Indians understood cosmology and the transformation of energy. Their predictions were accurate on the celestial movements and their influence on each other. The saints of India, through deep meditation, could visualise the cosmic formations and celestial movements.

Carl Sagan, the renowned cosmologist, once said

The Hindu religion is the only one of the world's great faiths dedicated to the idea that the Cosmos itself undergoes an immense, indeed an infinite number of deaths and rebirths. It is the only religion in which the time scales correspond to those of modern scientific cosmology. Its cycles run from our ordinary day and night to a day and night of Brahma, 8.64 billion years long. Longer than the age of the Earth or the Sun and about half the time since the Big Bang.

Konstantin Tsiolkovsky, considered to be the father of astronautics, envisioned humans conquering space and settling on Mars. He designed a rocket in 1886 for man to land on Mars and provided the rocket equation. Fascinated by his thoughts, the world has progressed to study the planets in the solar system and beyond to look for possible alternate abodes for humankind.

This book is an attempt to enlighten and inspire future generations about the understanding of the universe, with spacecraft and telescopes injected into space by rockets, the principle of rockets, new space technologies, and the future of the Earth and Sun.

2 Origin of the Universe

In this chapter, we will be discussing the very early understanding of the universe by ancient India, in Vedas, and the scientific postulations by the great saints, establishing India's heritage as the pioneer of visualising the universe. This chapter will also bring out in detail the new scientific thoughts of the west. We will also look at the evolution of the solar system and life on earth. Everything started with the inquisitive thinking of humans for exploration, investigation, and learning about the universe and the desire to acquire knowledge and skill on the wonders of the sky, and the use of space technology and rocketry.

VERSE FROM THE ANCIENT INDIA

RIG VEDA – Concept of Universe (8000–10,000 BCE)
Veda means wisdom or knowledge. The Four Vedas – Rig, Yajur, Sama, and Atharvana – represent glorious treasures in mythological evolution of cosmology and humanity. They are the embodiments of the immutable truth India stood for many years. According to tradition, they relate to the four ends of life. Rig Veda to *Moksha*, which refers to the air, the Spirit. The Yajur Veda to *Artha*, which implies to the element of Earth. The Sama Veda to *Kama,* which is indicative of water. The Atharvana Veda to *Dharma*, which is suggestive of the element fire.

Rigveda is the ancient of the four Veda books. It is the world's oldest religious writings, dated back to 8000 BCE, and some calculations date it back to 10,000 BCE.

Rig Veda verses aim to present the knowledge of natural, supernatural, and speculative thoughts in India. It was believed in the initial stage of the universe was in three-tier formation: Earth, Heaven, and in-between Antariksa (Space) in two hemispheres. The latter is inverted over the former. Indra rules the Heaven, and all Gods of nature live there. The Earthen hemisphere has a flat surface, overlooking Space and Heaven. Agni descended from Heaven to the eight corners of Earth as principal stations, defining the limits and shape of the universe.

Thou, Agni, shining in thy glory through the days, art

brought to life from out the waters, from the stone.

From out the forest trees and herbs that grow on ground

thou Sovran Lord men art generated pure.

Rig Veda defines the various elements of the universe such as Jyoti (Light), Tamas (Darkness), Aditya (Sun), Power of Agni, Moon (Soma), Heaven, Earth and Space, Time, Seasons, Life, Sound (Nada), Shakti (Power), Astronomy, Purusa-Sukta, mystic

DOI: 10.1201/9781003323396-2

symbol OM, and so on. Another explanation of the existence of dark matter and dark energy follows from the verses of Taittiriya Aranyaka that only one pada of the Lord is this manifested universe and the remaining three padas are outside it. According to the observations and realisations of ancient seers, nearly 80% of this universe is yet of unknown nature, which remains inactive during the process of creation (rather, evolution). This inactive matter may be treated as dark matter (or dark energy). In Nasadiya Sutra of Rigveda, there is also mention of this inactive matter:

Nothing existed prior to creation of universe, neither shape nor size nor kind nor even the light. The impenetrable darkness was filled with an unknown, unmovable fluid which was static and inactive (dark matter). In Samkhya philosophy also the dark matter was mentioned:

Naavastuni vastu siddhi

which means that only life and inert fluid (dark matter) existed before creation. Manusmrti also refers to this inactive (dark) matter:

In the beginning it was dark, inert, imperceptible, inexpressible unique matter which manifested in various forms at various times according to its own will.

Regarding the instantaneous creation of the dynamic universe, Rigveda says:

Yadkrandah prathamam prathamaam jaayamaana udyanat'smudraad ut vapureesaat,

Syenasya paksaa harnaasya baahu upastuyam mahijaatamam te arvan.

The above Sanskrit verse means that powerful light arose with the speed of a falcon, and several galaxies with central Hirangarbha came into existence from the invisible fluid (dark matter), which filled in the space and became the cause of Prakriti.

The second verse of Purush Sukta also gives an indication that the indestructible original element in the form of energy is effluent, and a small part of it is active in the creation of matter. In the third verse of this Sukta, there is a notable indication that only a small part of the energy, which exists in the form of universal energy, modifies and transforms to create this whole universe, and its remaining larger part remains untouched and inactive by this cycle of creation. **This sixth verse of Purush Sukta also says that very small part of infinite and un-destructive entity is engaged in controlling the whole cycle of creation, and the huge part remains inactive i.e., dark energy.**

The Rigveda contains the *Nasadiya sukta* hymn 10.129, which does not offer a cosmological theory, but asks cosmological questions about the nature of the universe and how it began:

Darkness there was at first, by darkness hidden.
Without distinctive marks, this all was water.
That which, becoming, by the void was covered.
That one by force of heat came into being.

Who really knows? Who will here proclaim it?
Whence was it produced? Whence is this creation?
Gods came afterwards, with the creation of this universe.
Who then knows whence it has arisen?

Whether God's will created it, or whether He was mute;
Perhaps it formed itself, or perhaps it did not.
Only He who is its overseer in highest heaven knows,
Only He knows, or perhaps He does not know.

(Reference to INTERNATIONAL JOURNAL OF SCIENTIFIC & TECHNOL-OGY RESEARCH VOLUME 6, ISSUE 01, JANUARY 2017 ISSN2277–8616 104 IJSTR©2017 www.ijstr.org Vedic Theory of Everything Subhendu Das)

Another explanation of **the existence of dark matter and dark energy** follows from the verses of Taittiriya Aranyaka that only one pada of the Lord is this manifested universe, and the remaining three padas are outside it. According to the observations and realisations of ancient seers, nearly 80% of this universe is yet of unknown nature, which remains inactive during the process of creation (rather, evolution). This inactive matter may be treated as dark matter (or dark energy).

Rig Veda's descriptions 10,000 years before coincide more or less with the later findings of Big Bang theory and the scientific debates.

Hindu Concept of Universe (2500–3000 BCE)
The Hindu concept of universe is quite different from Rigveda, though Vedas are claimed to be part of Hinduism. Most scholars believe Hinduism started somewhere around **2500 BCE** in the Indus Valley, but many Hindus argue that their faith is timeless and has always existed. Unlike other religions, Hinduism has no one founder but is instead a fusion of various beliefs.

HINDU PURANAS

In the Brahmanda Purana, as well as Bhagavata Purana (2.5), **14 lokas** (planes) are described, consisting of seven higher (Vyahrtis) and seven lower (Patalas) lokas. They are in layers, one above the other in sequence.

Higher lokas are **Satya-loka** (Brahma-loka), Tapa-loka, Jana-loka, Mahar-loka, Svar-loka (Svarga-loka or Indra-loka), Bhuvar loka, and **Bhu-loka (Earth plane).**

Lower lokas are Atala-loka, Vitala-loka, Sutala-loka, Talatala-loka, Mahatala-loka, Ras Atala-loka, and **Patala-loka.**

Hindu Cosmology upholds the idea that creation is timeless, having no beginning in time. Each creation is preceded by dissolution, and each dissolution is followed by creation. The whole cosmos exists in two states – the unmanifested or undifferentiated state and the manifested or differentiated state. This has been going on eternally. There are many universes – all follow the same rhythm, creation, and dissolution (the systole and diastole of the cosmic heart). **According to the Bhagavad Gita, this Srishti (creation) and pralaya (dissolution) recur at a period of 1000 mahayuga or 4.32 billion years.**

Puranas say that thousands of Brahmas have passed away! In other words, **this is not the first time the universe has been created.**

Shiva, the cosmic dancer, is the most perfect personification of the dynamic universe, from which emerges the general picture of organic, growing, and rhythmically moving cosmos.

The metaphor 'cosmic dance' has found its most profound beautiful expression in the image of Natraja (dancing Shiva), which depicts all life as part of a great rhythmical process of creation and annihilation as endless cycle. This cosmic dance of Shiva is the dance of creation and destruction involving the whole universe, the basis of all existence and all natural phenomena. It shows that the movement and rhythm are the essential properties of matter (and energy), where all matter (and energy) here on Earth or in outer space are involved in a continual cosmic dance. Jagat or Samsara, which are synonymous for the universe, means that one which is continuously moving. **Thus, the universe is continuously expanding in all directions.**

GREAT INDIAN SAINTS (SCIENTISTS)

In later years, there were great enunciations between 3000 BCE and 600 CE, much before the emergence of science in Europe. Everything that happened as scientific findings with experimental proof later had already been told by the great sages so many times before, though in different guises. The discoveries of sages came from an insight into the nature of reality through deep meditation and ascetic practises. For example, the realisation of Earth as a sphere by Acharya **Varahamitra** in 500 BCE and his calculation of the circumference of Earth at the equator as 40,350 km and diameter as 12,845 km (in equivalent units of that time), are very close to modern accepted data (40,075 km and 12,756 km, respectively). But such innovative thoughts were not communicated in time in an organised manner to the mankind. Above all, the most spectacular contribution was the concept of **zero,** without which modern computer technology would have been non-existent.

Albert Einstein said

We owe a lot to the Indians, who taught us how to count, without which no worthwhile scientific discovery could have been made!

Below are some discoveries by the great saints born in this country, to tell those who love purity of science and to acknowledge the Indian enunciations made well before the western scientists. (Reference: hindujagrruti.org/article/31.html)

Aryabhata (476 CE) – Master Astronomer and Mathematician Aryabhata, at the age of 23 in 499 CE, wrote a text on astronomy called "Aryabhatiyam." He formulated the process of calculating the motion of planets and the time of eclipses.

Aryabhata was the first to proclaim that the Earth rotates on its axis, orbits the sun, and is suspended in space – 1000 years before Copernicus published his heliocentric theory, *De revolutionibus orbium coelestium* in 1543 CE.

Aryabhata is also acknowledged for calculating π(Pi) to four decimal places: 3.1416 and the sine table in trigonometry. Centuries later, in 825 CE, the Arab mathematician, Mohammed Ibna Musa credited the value of Pi to the Indians.

Aryabhata was the forerunner and a colossus in the field of mathematics and astronomy.

Bhaskaracharya (1114–1183 BCE) – Genius in Algebra

Bhaskaracharya's work in algebra, arithmetic, and geometry was remarkable. His renowned mathematical works called "Lilavati" and "Bijaganita" are unparalleled and a memorial to his profound intelligence. Its translation in several languages of the world bear testimony to its eminence. In his treatise "Siddhant Shiromani", he wrote on planetary positions, eclipses, cosmography, mathematical techniques, and astronomical equipment. In the "Surya Siddhant", he makes a note on the force of gravity: "Objects fall on earth due to a force of attraction by the earth. Therefore, the earth, planets, constellations, moon, and sun are held in orbit due to this attraction." Bhaskaracharya was the first to discover gravity, many many years before Sir Isaac Newton (1642–1727 CE). He was the champion among mathematicians of ancient and medieval India. His works fired the imagination of Persian and European scholars, who through research on his works earned fame and popularity.

Acharya Kanad (600 BCE) – Founder of Atomic Theory

As the founder of "Vaisheshik Darshan" – one of six principal philosophies of India – Acharya Kanad was a genius in philosophy. He was the pioneer expounder of realism, law of causation, and the atomic theory. He classified all the objects of creation into nine elements, namely: earth, water, light, wind, ether, time, space, mind, and soul. He says, "Every object of creation is made of atoms which in turn connect with each other to form molecules." His statement ushered in the Atomic Theory for the first time ever in the world, nearly 2500 years before John Dalton. Kanad had also described the dimension and motion of atoms and their chemical reactions with each other. The eminent historian, T.N. Colebrook, has said, "Compared to the scientists of Europe, Kanad and other Indian scientists were the global masters of this field."

Acharya Kapil (3000 BCE) – Father of Cosmology

Acharya Kapil gifted the world with the Sankhya School of Thought, 5000 years before, enunciating the nature and principles of the ultimate Soul (Purusha), primal matter (Prakruti), and creation. His concept of transformation of energy and profound commentaries on atma, non-atma, and the subtle elements of the cosmos places him in an elite class of master achievers – incomparable to the discoveries of other cosmologists. On his assertion that Prakruti (visible and invisible matter) is the mother of cosmic creation and all energies, he contributed a new chapter in the science of cosmology. Because of his extrasensory observations and revelations on the secrets of creation and transformation of energy, he is recognised as the Father of Cosmology.

Acharya Bharadwaj (800 BCE) – Pioneer of Aviation Technology

Acharya Bharadwaj authored the "Yantra Sarvasva", which includes astonishing and outstanding discoveries in aviation science, space science, and flying machines. He described three categories of flying machines: 1) One that flies on Earth from one place to another; 2) one that travels from one planet to another; and 3) one that travels from one universe to another. His designs and descriptions have impressed and amazed aviation engineers of today. His brilliance in aviation technology is further reflected through techniques described by him:

1. **Profound Secret**: The technique to make a flying machine invisible (stealth) through the application of sunlight and wind force.
2. **Living Secret:** The technique to make an invisible space machine visible through the application of electrical force.
3. **Secret of Eavesdropping**: The technique to listen to a conversation in another plane.
4. **Visual Secrets**: The technique to see what's happening inside another plane.

Through his innovative and brilliant discoveries, Acharya Bharadwaj was recognised as the pioneer of aviation technology.

Much before 5000 years, Indians were the first to understand cosmology, the universe, celestial movements, and their influence on the life on Earth. With time, there is increasing awareness to the western world, of the existence of scientific base in ancient Indian Vedas.

The idea that ancient Indian texts have all the knowledge about the universe is a very popular one. One of the biggest arguments used to lend credence to that idea is how Hindu cosmology matches the current understanding of the universe. The primary evidence put forth for that argument is the number 4.32 billion years from Vishnu Purana, which is quite close to the current estimate *of the age of the Earth – 4.5 billion years. Carl Sagan acknowledged this fact.*

The universe is infinite in space and time and is infinitely heterogeneous. This means our Earth is not the only celestial body for life. Many such "earths" could be there in the universe, suitable for life. In our solar system, Earth with its Moon is an ideal planet suitable to harbour human life. The great enunciations of Indian saints, well before all other civilisations came to light after many centuries, are not known to the book writers of the west. They never mention India in their books, except a few.

 Patrick Brauckmann
@vonbrauckmann

Rig Veda is considered the oldest known text on the planet at pre 10,000 BC. It reminds us that the Dharmic culture of India knew the average distance from the Earth to both Moon and Sun were 108 times their diameters. How extraordinary and yet mysterious culture. 108.

EMERGING NEW SCIENTIFIC THOUGHTS ON THE UNIVERSE

Aristotle, the Greek philosopher, believed the universe had existed forever and there was no trace of change; hence, it was not created. His reasoning was that natural disasters had repeatedly set civilisation back to the beginning, and the cycle was going on. Something eternal is more perfect than something created to avoid invoking divine intervention. Conversely, those who believed the universe had a beginning, used it as an argument for the existence of God as the creator of the universe.

Then, how did it all start? How long will the universe be there? How long will the Sun shine? What will happen to humanity? What will happen to Earth? Will it become dry like Mars? Will the Moon and Mars become a factory for mining resources? Will they be launch pads for planetary missions? Only space exploration will answer these questions. NASA / JPL / ESA have done tremendous work on planetary explorations by sending spacecraft to far-off planets and orbiting tele-scopes to map the universe with its galaxies, stars, and black holes. The Hubble Telescope made remarkable contribution to understanding many unknowns. The arrival of the James Webb Telescope at L2 will throw light on mysteries of the universe. Let us learn about the theories and findings.

Modern understanding of the universe started in the 16th century with Nicholas Copernicus, Galileo Galilei, Johannes Kepler, Tycho Brahe, and Isaac Newton; and the 20th century has seen the emergence of great scientists and astronomers starting from Albert Einstein, Stephen Hawking, Edwin Hubble, and a few others providing clarity through their pronunciations. We will discuss some interesting areas for the understanding of the universe.

GENERAL THEORY OF RELATIVITY

The greatest scientist of the 20th century, Albert Einstein, published four articles in 1905 called the Annus Mirabilis papers. These four works, as below, changed the world's views on space, time, and matter:

1. The article that was pivotal for the development of quantum theory: *on a heuristic viewpoint concerning the production and transformation of light.*
2. The article that provided empirical evidence for the atomic theory: *on the motion of small particles in stationary liquid.*
3. The article that shed light on special relativity: *on electrodynamics of moving bodies.*
4. The groundbreaking article with the famous formula equating energy and mass $(E = mc^2)$: *does inertia of a body depend on its energy.*

He published all of the above in a single year, when he was just 26 years old.

In 1915, Albert Einstein introduced his revolutionary General Theory of Relativity to the world, which transformed the understanding of the universe, with space-time as a dynamic entity. He overthrew many assumptions underlying earlier physical theories and redefined the fundamental concepts of space, time, matter, energy, and also gravity.

Space-time is distorted by any matter that is contained in it. Matter tells space-time how to curve, and space-time tells matter how to move. General relativity predicts that even light is deflected by gravity – a prediction that has been confirmed by numerous astronomical observations. In addition, it predicts exotic phenomena like gravitational waves and black holes. Einstein's theory and his enunciations are getting proved, as follows.

1. **Time is the fourth dimension.**
2. **The speed of light remains constant.**
3. **The faster you move in space, the slower you move in time.**
4. **Time slows down around heavy objects.**
5. **Gravity is the curvature of space-time.**
6. **Gravity travels in the form of waves.**

In the early 1990s, Newton's law of universal gravitation published in 1687 had been accepted as a valid description of the gravitational force between masses. In Newton's model, gravity is the result of an attractive force between two massive objects.

Newton's law of universal gravitation states that every particle attracts every other particle in the universe with a force that is directly proportional to the product of their masses and inversely proportional to the square of the distance between their centres.

Experiments and observations show that Einstein's description of gravitation in relativity accounts for several effects that are unexplained by Newton's law, such as minute anomalies in the orbits of Mercury and other planets. General relativity also predicts novel effects of gravity, such as gravitational waves, gravitational lensing, and gravitational time dilation. Many of Einstein's predictions have been confirmed by experiments or observations, the most recent being gravitational waves.

General relativity provides the foundation for the current understanding of black holes, regions of space where the gravitational effect is strong enough that even light cannot escape. Their strong gravity is thought to be responsible for the intense radiation emitted by certain types of astronomical objects (such as active galactic nuclei or micro quasars). General relativity can be reconciled with the laws of quantum physics to produce a complete and self-consistent theory of **quantum gravity**. This is also part of the research framework of the standard Big Bang model of cosmology at CERN.

EINSTEIN'S EQUATIONS

Einstein's equations are the important part of general relativity. They provide a precise formulation of the relationship between space-time geometry and the properties of matter, using mathematics.

As has already been mentioned, the matter content of the spacetime defines another quantity, the energy–momentum tensor **T**, and the principle that **"space-time tells matter how to move, and matter tells spacetime how to curve"** means that these quantities must be related to each other (Figure 2.1).

The high-precision test of general relativity by the Cassini space probe (artist's impression): radio signals sent between the Earth and the probe (green wave) are delayed by the warping of spacetime (blue lines) due to the Sun's mass, as shown in Figure 2.2.

Einstein formulated this relation by using the Riemann curvature tensor and the metric to define another geometrical quantity **G**, now called the **Einstein tensor**, which describes the way spacetime is curved. *Einstein's equation* then states, that

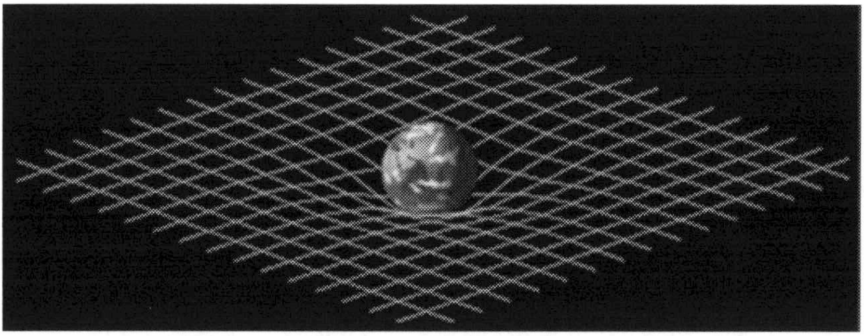

FIGURE 2.1   Schematic of spacetime curvature schematic.

FIGURE 2.2   Bending of radio signal.

$$\mathbf{G} = (8\pi G/c^4).\ \mathbf{T}$$

i.e., up to a constant multiple, the quantity **G** (which measures curvature) is equated with the quantity **T** (which measures matter content). Here, G is the gravitational constant of Newtonian gravity, and c is the speed of light from special relativity.

(In mathematics, a **tensor** is an algebraic object that describes a relationship between sets of algebraic objects related to a vector space).

PROVING EINSTEIN'S THEORY

- *Einstein's formulations on laws of physical forces ensure that all scientists will use the same laws, no matter how they are moving. The new law of gravity supersedes Newton's and now has become the intellectual challenge and best creation of humankind.*
- *The very important consequence of relativity is the relationship between mass and energy, $E = mc^2$. This is probably the only equation in physics recognised even by ordinary folk. Accelerating the particle to the speed of light, enormous energy can be produced. If the nucleus of a uranium atom fissions into two nuclei with less total mass, this will release a huge amount of energy.*
- *Stephen Hawking asserted that Einstein's theory of relativity, established through several experiments, proved that time and space are inextricably interconnected. One cannot curve space without involving time. Thus, time has a shape. Because gravity is attractive, matter always warps spacetime so that light rays bend toward each other.*

PROVED EINSTEIN'S THEORIES

Gravitational lensing, gravitational waves, time dilation, and black holes have been proved. His theories on white holes and worm holes are yet to be proved Figure 2.3.

FIGURE 2.3 Proving Einstein's predictions on the general theory of relativity.

GRAVITATIONAL LENSING

A gravitational lensing is a distribution of matter (such as a cluster of galaxies) between a distant light source and an observer that is capable of bending the light from the source as the light travels towards the observer.

A quasar, also known as a quasi-stellar object, is an extremely luminous active galactic nucleus, in which a super massive black hole ranging from millions to billions of times the mass of the Sun is surrounded by a gaseous accretion disk.

The Einstein Cross: *four images of the same distant quasar*, produced by a gravitational lens. Since the light of a very distant object is deflected in a gravitational field across many galaxies, it is possible for the light to reach an observer along two or more paths. As a result, the observer on Earth will see one astronomical object in multiple places in the night sky. This kind of gravitational effect is called gravitational lensing.

GRAVITATIONAL WAVES

Just like an ocean wave is a slosh of water and an acoustic wave is a movement of air, gravitational waves are likewise ripples in the fabric of space-time. If you imagine the world around you covered with 3D grid lines, the warping and stretching of those physical coordinates would be a gravitational wave.

Gravitational waves, a direct consequence of Einstein's theory, are distortions of geometry that propagate at the speed of light, as ripples in space-time. In February 2016, the Advanced LIGO team announced that they had detected gravitational waves from merger of two stars. Such pairs of stars orbit each other and, as they do so, gradually lose energy by emitting gravitational waves. Gravitational wave observations can be used to obtain information about compact objects, such as neutron stars and black holes, and also to probe the state of the early universe fractions of a second after the Big Bang.

The detected waves were caused by two black holes spiralling in toward one another, and eventually crashing into one another. Scientists have released the largest catalogue of 35 gravitational wave events from the 90 detections to date, shedding new light on interactions between the most massive objects in the universe, black holes, and neutron stars.

COSMOLOGY

General relativity can be applied to the universe as a whole. All current observations suggest that, on average, the structure of the cosmos should be approximately the same, regardless of an observer's location or direction of observation. The universe is approximately homogeneous and isotropic. Such a comparatively simple universe can be described by simple solutions of Einstein's equations. The current cosmological models of the universe are obtained by combining these simple solutions to general relativity with theories describing the properties of the universe's matter content, namely thermodynamics, nuclear physics, and particle physics. According to these models, our present universe emerged from an extremely dense high-temperature state – the Big Bang – roughly 13.8 billion years ago and has been expanding ever since.

Big Bang Theory

The Big Bang theory is the prevailing cosmological model explaining the existence of the observable universe from the earliest known periods through its subsequent large-scale evolution. **The model describes how the universe expanded from an initial state of high density and temperature big bang, and inflation started at 10^{-36} seconds. It offers a comprehensive explanation for a broad range of observed phenomena, including the abundance of chemical elements such as deuterium, helium, and others, the cosmic microwave background (CMB) radiation, and large-scale structure.**

According to estimation of this theory, space and time emerged together 13.799 ± 0.021 billion years ago, and the energy and matter initially present have become less dense as the universe expanded from the glow, as described in Figure 2.4.

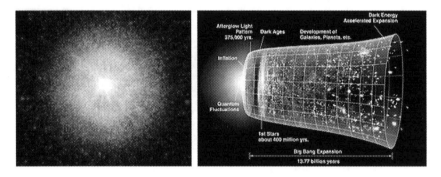

FIGURE 2.4   BIG BANG and timeline of the metric expansion of space, where space, including hypothetical non-observable portions of the universe, is represented at each time by the circular sections. On the left, the dramatic expansion occurs in the inflationary epoch; and at the centre, the expansion accelerates (artist's concept; not to scale).

Fred Hoyle, an English cosmologist, was the first to call this process the Big Bang, in 1949. It is believed that this incredibly dense point of primitive matter/energy exploded. Within seconds, the fireball ejected matter/energy at velocities approaching the speed of light or more. At some later time, energy and matter began to split apart and become separate entities.

CERN – Finding the GOD Particle

CERN, the European Organization for Nuclear Research, is one of the world's largest and most respected centres for scientific research. Its business is fundamental physics, finding out what the universe is made up of and how it works. When I went to CERN with Dr. APJ Abdul Kalam in 2005, I saw at the entrance a big statue of Nataraja (Shiva), symbolising Shiva's cosmic dance of creation.

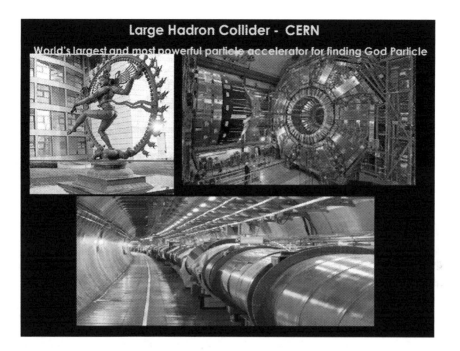

FIGURE 2.5   Large Hadron Collider – CERN.

Scientists from many countries, including India, are participating in the research. CERN is also the birthplace of the World Wide Web.

Researchers at CERN are using 27 km circular accelerator called the **Large Hadron Collider (LHC)** to accelerate subatomic particles called protons to close to the speed of light (Figure 2.5). This is to understand the formation of the universe and the particles that moved in the instants after the Big Bang. **It is believed that these Big Bang particles should have speed more than the speed of light.**

When heavy nuclei smash into one another in the LHC, the hundreds of protons and neutrons that make up the nuclei release a large fraction of their energy into a tiny volume, creating a fireball of quarks and gluons. These tiny bits of quark–gluon plasma only exist for fleeting moments. By studying the particles produced in the collisions – before, during, and after the plasma is created – researchers can study the plasma from the moment it is produced to the moment it cools down and gives way to a state in which composite particles called hadrons can form. However, the plasma cannot be observed directly. Its presence and properties are deduced from the experimental signatures it leaves on the particles that are produced in the collisions and their comparison with theoretical models.

Universe after BIG BANG

After the Big Bang, the universe got shaped at 10^{-32} seconds, basic elements got evolved at one micro second, and stars and galaxies formed after 200 million years (Figure 2.6). The existence of an expanding universe implies that the cosmos has

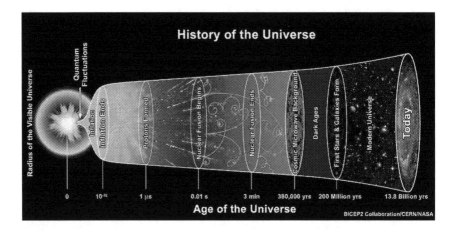

FIGURE 2.6   Illustration of the history of the universe. 0, 10^{-32} and about one microsecond from the Big Bang, protons formed from the quark–gluon plasma. (Image: BICEP2 Collaboration/CERN/NASA).

evolved from a dense concentration of matter into the present broadly spread distribution of galaxies. Most cosmologists and theoretical physicists endorse this theory because of overwhelming evidence. Still, there are prominent cosmologists who question the Big Bang theory as well as a single universe.

The **Higgs boson** (named after Prof. Higgs and Dr. S N Bose – famous for Bose–Einstein condensate) is the particle associated with the Higgs field, an energy field that transmits mass to the things that travel through it. When they collide, they create super-high-energy mashups that spew out subatomic particles. From time to time, a Higgs boson might be one of those particles.

Neutrinos – ghostly subatomic particles – during initial tests, might have been observed travelling a fraction of second faster than the speed of light, CERN scientists announced.

"If this is true, it would rock the foundations of physics," said Stephen Parke, head of the theoretical physics department at the Fermilab near Chicago, immediately after the initial claim. But later tests confirmed, **the neutrino particles also travel at the speed of light.** This proves the cardinal rule of physics established by Albert Einstein nearly a century ago.

Experiments need to be continued to find out that GOD particle having speed more than that of light, to disprove Einstein.

The Redshift – Our Expanding Universe

In 1929, astronomer **Edwin Hubble**, working at the Mount Wilson Observatory in California, announced that all the galaxies he had observed were receding from Earth and from each other, at speeds of up to several thousand kilometres per second (more than speed of light). Hubble found that stars are not uniformly distributed throughout space but are gathered, together in vast collections called galaxies. By measuring the light from galaxies, Hubble could determine their velocities. The universe was

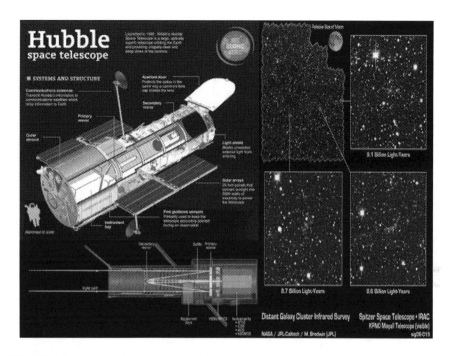

FIGURE 2.7 Galaxy cluster seen by Hubble.

changing with time, and it was expanding. **The distance between distant galaxies was increasing with time** (Figure 2.7).

To clock the speeds of these galaxies, Hubble took advantage of the Doppler Effect. This phenomenon occurs when a source of waves, such as light or sound, is moving with respect to an observer or listener. If the source is moving away, the waves drop in frequency: sound becomes lower in pitch, and light tends to shift toward the red end of the spectrum of the light from the galaxies.

Hubble used a spectroscope, a device that analyses the different frequencies present in light. He discovered that the light from galaxies far off in space was shifted down toward the red end of the spectrum. Where in the sky each galaxy lay didn't matter – all were redshifted. Hubble explained this shift by concluding that the galaxies were in motion, whizzing away from Earth. The greater the redshift, Hubble assumed, the greater the galaxy's speed.

The observations made by Hubble on redshift demonstrated that distant galaxies together are moving away from the Earth with speeds that are proportional to the distances from us. That means these galaxies must have formed together at the same time and are moving together.

FUNDAMENTAL PHYSICAL FORCES

Physical forces are acting all around us in any routine activity, including a rocket launch into space. But we don't realise all the forces that we experience every day. There are four fundamental forces:

- Gravity (works over large distances)
- Electromagnetism (holds atoms together)
- The weak force (interactions of electrons)
- The strong force (holds nuclei together)

These are called the four fundamental forces of nature, and they govern everything that happens in the universe.

GRAVITY

Gravity is the attraction between two objects that have mass or energy, like the Moon causing ocean tides on earth. Gravity is probably the most intuitive and familiar of the fundamental forces, but it's also been one of the most challenging to explain.

Isaac Newton described gravity as a literal attraction between two objects. Newton's law of gravity states that the gravitational force between two bodies is proportional to the product of their masses and inversely proportional to the square of the distance between them.

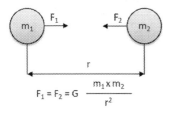

where,

F is the gravitational force

G is the universal gravitational constant, $G = 6.67 \times 10^{-11}$ Newtons $kg^{-2}m^2$

m1 and m2 are the masses of the objects in consideration

r is the distance between the centres of the two objects in consideration

Gravity is what keeps the planets in orbit around the Sun, and the Moon around the Earth.

Centuries later, Albert Einstein suggested, through his theory of general relativity, **that gravity is not an attraction or a force. Instead, it's a consequence of objects bending space-time.** Though gravity holds planets, the solar system, stars, and galaxies together, it turns out to be the weakest of the fundamental forces, especially at the molecular and atomic scales.

ELECTROMAGNETIC FORCE

The electromagnetic force, also called the Lorentz force, acts between charged particles, like negatively charged electrons and positively charged protons. Opposite charges attract one another, while like charges repel. The greater the charge, the greater the force. And much like gravity, this force can be felt from an infinite distance.

As its name indicates, the electromagnetic force consists of two parts: the electric force and the magnetic force. The electric component acts between charged particles whether they are moving or stationary, creating a field by which the charges can influence each other. But once set into motion, those charged particles begin to display the second component, the magnetic force. The particles create a magnetic field around them as they move.

The electromagnetic force is responsible for some of the most experienced phenomena: friction, elasticity, the normal force, and the force holding solids together in a given shape. These actions can occur because of charged (or neutralised) particles interacting with one another.

THE WEAK NUCLEAR FORCE

The weak force, also called the weak nuclear interaction, is responsible for particle decay. This force causes radioactivity and plays vital role in the formation of the elements in stars and the early universe. This is the literal change of one type of subatomic particle into another. So, for example, a neutrino that strays close to a neutron can turn the neutron into a proton while the neutrino becomes an electron.

Physicists describe this interaction through the exchange of force-carrying particles called **bosons.** Specific kinds of bosons are responsible for the weak force, electromagnetic force, and strong force. In the weak force, the bosons are charged particles called W and Z bosons. When subatomic particles such as protons, neutrons, and electrons come within 10^{-18} m, or 0.1% of the diameter of a proton, of one another, they can exchange these bosons. As a result, the subatomic particles decay into new particles, according to Georgia State University's Hyper Physics website.

The weak force is critical for the nuclear fusion reactions that power the sun and produce the energy needed for most life forms here on Earth. It's also why archaeologists can use carbon-14 to date ancient bone, wood, and other formerly living artefacts. Carbon-14 has six protons and eight neutrons; one of those neutrons decays into a proton to make nitrogen-14, which has seven protons and seven neutrons. This decay happens at a predictable rate, allowing scientists to determine how old such artefacts are.

THE STRONG NUCLEAR FORCE

It is the strongest of the four fundamental forces of nature. It is 6000 trillion trillion trillion (that's 39 zeros after 6!) times stronger than the force of gravity. That is because it binds the fundamental particles of matter together to form larger particles. It holds together the quarks that make up protons and neutrons, and part of the strong force also keeps the protons and neutrons of an atom's nucleus together.

UNIFYING NATURE

The outstanding question of the fundamental forces is whether they are manifestations of just a single great force of the universe. If so, each of them should be able to merge with the others, and there's already evidence that they can. The theory of everything is a theoretical framework that could explain the entire universe.

Physicists, however, have found it difficult to merge the microscopic world with the macroscopic one. At large and especially astronomical scales, gravity dominates and is best described by Einstein's theory of general relativity. But at molecular, atomic, or subatomic scales, quantum mechanics best describes the natural world. And so far, no one has come up with a good way to merge those two worlds.

BRANEWORLD

The forces of electromagnetism, radioactivity, and nuclear interactions are confined to a three-dimensional braneworld, while gravity acts in all dimensions of space, and so it is found to be much weaker than other forces on the braneworlds. (Ref. John D Barrow – The book of universes) as shown in Figure 2.8.

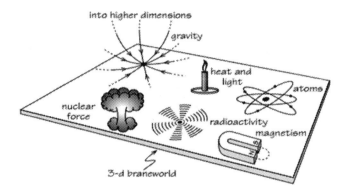

FIGURE 2.8   3-D braneworld.

MUONS

The investigation, based at the Fermilab or Fermi National Accelerator Laboratory in Batavia, Illinois, looks for signs of new phenomena in physics by examining the behaviour of sub-atomic particles, as described in CERN's site, also known as **'muons.'**

There are building blocks of the world that are even tinier than the atom. Some of these sub-atomic particles comprise even smaller constituents, while others cannot be broken down into any other fundamental particles. The muon is one of these essential particles. It is akin to the electron, although it's more than 200 times heavier.

The Muon g-2 experiment, as described by Fermilab, encompasses sending the particles around a 14 m ring and then applying a magnetic field. **The physicists discovered that muons wobbled at a speedier rate than expected. This might be driven by a force of nature that's totally new to science.**

STRING THEORY

In particle physics string theory is a theoretical framework in which the point-like particles are replaced by one-dimensional objects called strings. **It is to say that**

all matter and energy in the universe are composed of tiny strings. These strings may have ends or they may join up with themselves in closed loops. String theory describes how these strings propagate through space and interact with each other. On distance scales larger than the string scale, a string looks just like an ordinary particle, with its mass, charge, and other properties determined by the vibrational state of the string.

String theory is an attempt to unite the two pillars of 20th-century physics – quantum mechanics and Albert Einstein's theory of relativity – with an overarching framework that can explain all of physical reality. It tries to do so by postulating that particle is one-dimensional, string-like entities whose vibrations determine the particles' properties, such as their mass and charge. Many scientists believe in string theory because of its mathematical beauty. The equations of string theory are described as elegant, and its descriptions of the physical world are considered extremely satisfying. In string theory, one of the many vibrational states of the string corresponds to the graviton, a quantum mechanical particle that carries gravitational force. **Thus, string theory is a theory of quantum gravity.**

STELLAR EVOLUTION – LIFECYCLE OF A STAR

Stars are formed from collapsing clouds of gas and dust, often called nebulae or molecular clouds. Over the course of millions of years, these proto stars settle down into a state of equilibrium.

Stellar evolution is the process by which a star changes over the course of time. Depending on the mass of the star, its lifetime can range from a few million years for the most massive to trillions of years.

Nuclear fusion powers a star for most of its existence. Initially the energy is generated by the fusion of hydrogen atoms at the core of the main-sequence star. Later, as the preponderance of atoms at the core becomes helium, stars like the Sun begin to fuse hydrogen along a spherical shell surrounding the core. This process causes the star to gradually grow in size, passing through the subgiant stage until it reaches the red-giant phase. Stars with at least half the mass of the Sun can also begin to generate energy through the fusion of helium at their core, whereas more-massive stars can fuse heavier elements along a series of concentric shells. Once a star like the Sun has exhausted its nuclear fuel, its core collapses into a dense **white dwarf** and the outer layers are expelled as a planetary nebula. Stars with around ten or more times the mass of the Sun can explode in a supernova as their inert iron cores collapse into an extremely dense **neutron star or black hole.** Although the universe is not old enough for any of the smallest red dwarfs to have reached the end of their existence, stellar models suggest they will slowly become brighter and hotter before running out of hydrogen fuel and becoming low-mass white dwarfs as shown in Figure 2.9.

In 1931, Subrahmanyam Chandrasekhar calculated, using special relativity, that a non-rotating body of electron-degenerate matter above a certain limiting mass (now called the Chandrasekhar limit at 1.4 $M_\odot$) has no stable solutions. ($M_\odot$ is the mass of Sun (1.98847±0.00007) × 10^{30} kg). When the mass exceeds 1.4 $M_\odot$ electron degeneracy is no longer strong enough to resist the pull of gravity, and the white dwarf abruptly collapses into a neutron star.

Chandra explained the nature of white dwarfs (and won the Nobel Prize for his work).

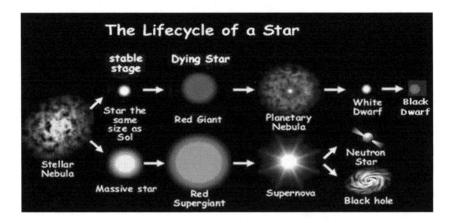

FIGURE 2.9   The life of a star.

Image Courtesy: schoolsobservatory.org.

BLACK HOLE

A black hole is a cosmic body of extremely intense gravity from which even light cannot escape. Black holes usually cannot be observed directly, but they can be "observed" by the effects of their enormous gravitational fields on nearby matter. The singularity constitutes the centre of a black hole, hidden by the object's "surface," the event horizon. Inside the event horizon, **the escape velocity exceeds the speed of light so that not even rays of light can escape into space.**

A black hole can be formed by the death of a massive star. At the end of a massive star's life, the core becomes unstable and collapses in upon itself, and the star's outer layers are blown away. The crushing weight of constituent matter falling in from all sides compresses the dying star to a point of zero volume and infinite density, called the singularity.

Once a black hole forms, it continuously grows by absorbing all the mass from its vicinity and at certain times, can even merge with other black holes. Scientists hypothesise that at the centre of any galaxy is a supermassive black hole.

Albert Einstein first predicted the existence of black holes in 1916, with his general theory of relativity. The term "black hole" was coined many years later in 1967 by American astronomer John Wheeler. After decades of black holes being known only as theoretical objects, the first physical black hole ever discovered was spotted in 1971.

Then, in 2019 the Event Horizon Telescope (EHT) collaboration released the first image ever recorded of a black hole. The EHT saw the black hole in the centre of galaxy M87 while the telescope was examining the event horizon, or the area past which nothing can escape from a black hole. The image maps the sudden loss of photons (particles of light). It also opens a whole new area of research in black holes, now that astronomers know when a star burns through the last of its fuel, the object may collapse, or fall into itself. For smaller stars (those up to about three times the sun's mass), the new core will become a neutron star or a white dwarf. But when a larger star collapses, it continues to compress and creates a stellar black hole.

In addition, the quantum field theory predicts that black holes also emit a type of radiation called the Hawking Radiation. This is defined as a black body with a temperature inversely proportional to its own mass.

Black holes are categorised into four **types.** The first is the **supermassive black hole or SMBH.** This is the largest type amounting to an unmeasurable number of solar masses. The second type is the **intermediate mass black hole,** which is a hypothetical class with a mass ranging from 100 to 1,000,000 solar masses. Although the existence of intermediate mass black holes has still yet to be proven, there is indirect evidence that they exist using various positions from known stars. The third type, known as the **stellar black hole or stellar mass black hole** is formed when a massive star collapses. Their masses range from five to 100 solar masses and can be observed as either a gamma ray burst or a hypernova explosion. These types of black holes are also called collapsars. The last type is known as a **micro black hole, a mini black hole, or a quantum mechanical black hole**. They were introduced by Stephen Hawking in 1971.

There are three (3) major black holes observed near our galaxy.

A0620-00. This is a binary star system that belongs to the Monoceros constellation. This system consists of two main objects, a sequence star and an unknown mass where scientist believe there to be a stellar mass black hole. This system is approximately three thousand (3000) light years away.

Cygnus X-1. This is a galactic X-ray source type of system found in the constellation of Cygnus and is widely accepted by scientists to be a black hole. Discovered in 1964, this system is the most studied astronomical objects in space, estimated to have a mass of 15 times that of our Sun. This same system also belongs to a stellar association referred to as the Cygnus OB3. This means that the Cygnus X-1 is about five million years old and came from a progenitor star with more than 40 solar masses.

V404 Cygni. This is a binary system and a micro quasar consisting of a black hole bearing a mass of 12. It also has a K companion star with a smaller mass than that of our Sun. The black hole and the star orbit each other at close range. Because of the intense gravity of the black hole (and their proximity as well), the star loses mass to the black hole's accretion disk.

There are five **main parts of a black hole**. The first is the *event horizon,* which is a black hole's defining feature. This is the boundary where matter and electromagnetic radiation can only pass into the mass of the black hole and can no longer escape. The second is the *singularity,* which is the eye of the black hole. This is a region when the curvature becomes infinite. The third is the *photon sphere,* which is a spherical boundary (with no thickness), where photons move perpendicularly to the sphere, which is trapped in an elliptical orbit with respect to the black hole. The next part is the *ergo sphere*; this is the area where objects are always on the move because of a phenomenon called frame-dragging. The fifth and final part of a black hole is the *innermost stable circular orbit* or ISCO wherein particles stably orbit at various distances from a centre object.

According to **Stephen Hawking**, under general conditions, the total area of a black hole never decreases even after it absorbs mass. This hypothesis is now called the **second law of black hole mechanics,** which is quite similar to the second law of thermodynamics, which states that the total sum of entropy of any system can never decrease. The link of these two laws was further supplemented by the fact that black holes can radiate blackbody radiation at a particular temperature. **This event was based on the quantum field theory discovered by Stephen Hawking.**

Supermassive black holes could arise from large clusters of dark matter. This is a substance that we can observe through its gravitational effect on other objects. However, we don't know what dark matter is composed of because it does not emit light and cannot be directly observed.

The supermassive black hole at the core of supergiant elliptical galaxy Messier 87, with a mass about seven billion times that of the Sun, as depicted in the first false-colour image in radio waves released by the Event Horizon Telescope (10 April 2019). Visible are the crescent-shaped emission ring and central shadow, which are gravitationally magnified views of the black hole's photon ring and the photon capture zone of its event horizon. The crescent shape arises from the black hole's rotation and relativistic beaming; the shadow is about 2.6 times the diameter of the event horizon (Figure 2.10).

FIGURE 2.10 M 87 black hole.

Image Credit: phys.org **kids.nationalgeographic.com.**

Dark Energy and Dark Matter

What the Universe Is Made Of

A major part of the universe is covered by dark energy 74% and dark matter 21% (as of 2021). All stars, planets, asteroids, and gases all put together belong to 5% of the universe. In the future, the proportion of dark energy will increase, as the universe is expanding. (Some scientists claim the dark energy is 72% and dark matter is 23%), as given in Figure 2.11.

Out of the remaining 5%, hydrogen and helium occupy 4%, stars 0.5%, neutrino 0.3%, and heavy elements like planets (trillions of them) occupy only 0.03% of the universe. Think of Earth; it is an insignificant dot in the universe.

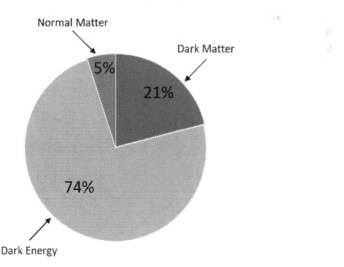

FIGURE 2.11 Universe composition.

Dark Energy

As we discussed in red shift, more distant galaxies are moving away from us at greater speed, and the universe is expanding. The acceleration implies there is a strong repulsive force (anti-gravity), which we call **dark energy**. It is a vacuum energy. The universe has 74% of active dark energy, and it will be increasing. This hypothetical form of dominantly occupied energy is responsible for the expanding universe.

Background to Dark Energy

Gravity's most familiar characteristic is that it pulls, not pushes. Einstein's theory of gravity introduced some remarkable ideas: black holes, curved space, and repulsive gravity. Even in Einstein's theory, matter always pulls; the repulsive aspect of gravity – a force that pushes matter apart – only arises in extraordinary circumstances.

Einstein happened to dwell upon this new feature in his theory – the repulsive force – when he introduced the "cosmological constant" (usually symbolised by lambda, Λ). – a sort of "fudge factor." He balanced the repulsive gravity of his cosmological constant against the attractive gravity of ordinary matter to create a static model of the universe. He discarded the cosmological constant when Hubble discovered that the universe is not static at all, but is expanding, as we saw in Redshift.

The discovery in 1998 that the expansion of the universe is speeding up and not slowing down, revived interest in repulsive gravity. Current hypotheses propose dark energy might emerge from the bubbling of empty space, a small effect that is also widespread, making it powerful enough to drag apart clusters of galaxies without ripping them apart from within. It does not have any local gravitational effects, but rather a global effect on the universe.

An attempt was made to measure the rate of expansion of the universe and its acceleration by observations based on the Hubble law. These measurements, together with other scientific data, have confirmed the existence of dark energy, and provide an estimate of just how much of this mysterious substance exists. **While there is no full consensus as to what the dark energy is, this mysterious force is extremely important to accelerate the expansion of the universe.**

Burst of Supernova

Scientists studied a specific case of Ia supernova explosion to understand the mysterious energy, which happened ten billion years ago, when the universe was in its infancy. The titanic supernova explosion unleashed streams of brilliant light. As the blazing light travelled through space and across billions of years, journeying toward Earth, it formed.

Astronomers using the Hubble telescope to hunt for distant supernovas caught it while taking a second look at the Hubble Deep Field. Astronomers could compare and standardise these light intensity curves, providing a tool to study the distant cosmos. The curves had the same shape because they start with the same type of source. Astronomers knew how much energy a **type Ia supernova** gives off during a blast, so the observed brightness gave them that supernova's distance as ten billion light years from Earth (Figure 2.12).

This stellar explosion is extraordinary not only because of its tremendous distance, but also because its discovery greatly bolsters the case for the existence of a mysterious form of "dark energy" pervading the universe.

In the late 1980s, Lawrence Berkeley National Laboratory astronomer Saul Perlmutter created the Supernova Cosmology Project to use these blasts to track

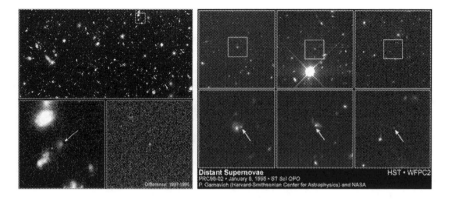

FIGURE 2.12   Distant supernova.

Credit: P. Garnavich (Harvard-Smithsonian Center for Astrophysics) and the High-z Supernova Search Team and NASA.

how our universe is expanding. Perlmutter and Schmidt won the 2011 Nobel Prize in Physics for their ground-breaking work, along with Adam Riess, who led the High-Z Supernova Search Team's analysis. The discovery of accelerating expansion was hailed as the discovery of the decade.

The Hubble discovery also reinforces the startling idea that the universe only recently began speeding up, **expanding more quickly than in the past**.

In the next one billion years, the universe will witness a faster turbulent expansion due to vigorous influence of dark energy, as shown in Figure 2.13.

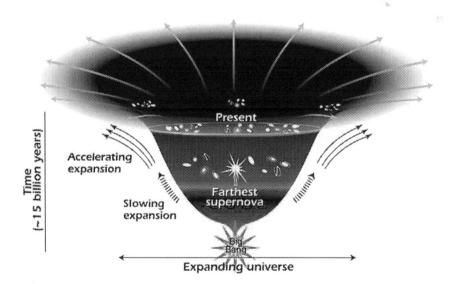

FIGURE 2.13   Accelerated expansion of the universe due to dark energy.

Credit: NASA/STSci/Ann Field.

This diagram reveals changes in the rate of expansion since the universe's birth 14 billion years ago. The shallower the curve, the faster the rate of expansion. The curve changes noticeably about seven billion years ago, when objects in the universe began flying apart at a faster rate. Astronomers theorise that the faster expansion rate is due to a mysterious, dark force that is pulling galaxies apart.

THE DYNAMICS OF UNIVERSE – IT WILL NEVER SHRINK

The work of Perlmutter, Schmidt, and Riess was the culmination of a decade of research slowly uncovering clues to an accelerating universe, and they provided a way to solve several cosmic mysteries.

The leading theory is vacuum energy, also known as the cosmological constant, which is basically the idea that empty space is not actually empty.

"Even if you take all of the matter out of space, the light, any neutrino, any particles whatsoever, in complete nothingness, you're never left with complete nothingness," opined Tamara Davis, University of Queensland. According to quantum physics, the virtual particles in vacuum could pop in and out of existence and create a type of negative pressure – **they push instead of pull.**

NEW DIMENSION

Another possibility is that **dark energy is perhaps a "quintessence", a yet undiscovered force of nature**. A universe with a variable density of matter would expand at different rates in different places, possibly producing an illusion of accelerated expansion.

Astronomers hope soon to know more from the Cerro Tololo Inter-American observatory in the Chilean Andes and from James Webb space telescope and to come to grips with this cosmic mystery by mapping out distant cosmic galaxies and the history of expansion of the universe over the almost 14 billion years since the Big Bang.

DARK MATTER

In 1933, astronomer **Fritz Zwicky** noticed that galaxies in clusters were moving at greater speeds, holding the cluster together. Later in the 1950s, astronomer **Vera Rubin** found that galaxies were spinning too fast to hold together. The astronomers concluded that galaxies and the clusters both behave as if more mass is present than we can find in stars, gas, and dust. **That missing mass is dark matter,** which cannot be seen. All galaxies are embedded in clouds of dark matter.

Dark matter is composed of particles that do not absorb, reflect, or emit light, so they cannot be detected by observing electromagnetic radiation. Dark matter is material that cannot be seen directly. We know that dark matter exists because of the effect it has on objects that we can observe directly. Approximately one quarter of the universe is dark matter.

In brief, this is a form of matter that only seems to interact through gravitation. **Several strands of evidence lend support to the existence of this form of matter**. For example, we can see a gravitational pull of galaxy clusters and other structures in the universe. We know that the matter in such structures is not enough to hold them together by gravity alone, **meaning some additional invisible matter must be present to make them spin at the speeds observed**.

Proving the existence of dark matter

If scientists can't see dark matter, how do they know it exists?

Scientists calculate the mass of large objects in space by studying their motion. They found the stars in the galaxy travelled at the same velocity, indicating the galaxies contained more mass than could be seen. Studies of the gas within elliptical galaxies also indicated more mass than found in visible objects. This mass is the dark matter.

Dark matter is likely to be causing the arcs seen around the central galaxies in Figure 2.14, making a smiling face.

The **rotation curve** of a disc galaxy (also called a **velocity curve**) is a plot of the orbital speeds of visible stars or gas in that galaxy versus their radial distance from that galaxy's centre. It is typically rendered graphically as a plot, and the data observed from each side of a spiral galaxy are generally asymmetric, so the data from each side are averaged to create the curve. A significant discrepancy exists between the experimental curves observed, and a curve derived by applying gravity theory to the matter observed in a galaxy. Theories involving dark matter are the main postulated solutions to account for the variance.

FIGURE 2.14   Cosmic smiley face in space galaxy rotation curve.

Image Credit: NASA/ESO; Acknowledgement: Judy Schmidt.

Albert Einstein showed that massive objects in the universe bend and distort light, allowing them to be used as lenses. By studying how light is distorted by galaxy clusters, astronomers have been able to create a map of dark matter in the universe. **All these methods provide a strong indication that most of the matter in the universe is something yet unseen.**

Scientists have discovered that some small fraction of dark matter is made of neutrinos – tiny, fast-moving particles that don't really interact with normal matter, like the Gran Sasso National Laboratory in Italy and the Deep Underground Science and Engineering Laboratory in South Dakota, are trying to detect dark matter particles directly, when they crash into atoms in cryogenically cooled tanks filled with liquefied gases. So far, they haven't managed to capture a dark matter particle in action. But researchers are taking dark matter into account when they think about how the universe evolves.

Dark matter works like an attractive force – a kind of cosmic cement that holds our universe together. This is because dark matter does interact with gravity, but it doesn't reflect, absorb, or emit light. The dynamics of galaxy clusters, Galactic rotation curves, the Cosmic Microwave Background, the Bullet Cluster, and Largescale Structure formation provide evidence for the existence of dark matter.

MULTI-UNIVERSE

Multiple universes have been hypothesised in cosmology, physics, astronomy, religion, philosophy, transpersonal psychology, music, and all kinds of literature, particularly in science fiction, comic books, and fantasy. In these contexts, parallel universes are also called "alternate universes".

Nobel Laurate Roger Penrose in his book "Cycles of Time" provides a completely different view of the origin of the universe, questioning what came before the Big Bang. His view is that the accelerating and expanding universe is heading for another Big Bang. John D. Barrow in his book on universes has debated the various multiverse theories over time. Prominent physicists are divided about whether any other universes exist outside of our own.

Some physicists say the multiverse is not a legitimate topic of scientific inquiry. Some have argued that the multiverse is a philosophical notion rather than a scientific hypothesis because it cannot be empirically falsified.

In 2007, Nobel laureate Steven Weinberg suggested that if the multiverse existed,

> the hope of finding a rational explanation for the precise values of quark masses and other constants of the standard model that we observe in our Big Bang is doomed, for their values would be an accident of the particular part of the multiverse in which we live.

Around 2010, scientists such as Stephen M. Feeney analysed Wilkinson Microwave Anisotropy Probe (WMAP) data and claimed to find evidence suggesting that this universe collided with another (parallel) universe in the distant past. However, a more thorough analysis of data from the WMAP and from the Planck satellite, which has a resolution three times higher than WMAP,

did not reveal any statistically significant evidence of such a bubble universe collision. In addition, there was no evidence of any gravitational pull of other universes on ours.

MILKYWAY AND SOLAR SYSTEM

The Milky Way is our galaxy, 13.2 billion years old, residing in a knot of two dozen galaxies. This galaxy contains our solar system, with the name describing the galaxy's appearance from Earth. It is a hazy band of light seen in the night sky formed from stars that cannot be individually distinguished by the naked eye. From Earth, the Milky Way appears as a band because its disk-shaped structure is viewed from within. Galileo Galilei first resolved the band of light into individual stars with his telescope in 1610. Until the early 1920s, most astronomers thought that the Milky Way contained all the stars in the universe. Following the 1920 Great Debate between the astronomers Harlow Shapley and Heber Curtis, observations by Edwin Hubble showed that the Milky Way is just one of many galaxies.

The Milky Way is a barred spiral galaxy with an estimated visible diameter of 150–200,000 light years, an increase from traditional estimates of 100,000 light years. The Milky Way has several satellite galaxies and is part of the local group of galaxies, which form part of the Virgo Supercluster, which is itself a component of the Laniakea Supercluster.

It is estimated that Milky Way galaxy contains 100–400 billion stars and at least that number of planets. The solar system is located at a radius of about 27,000 light years from the Galactic Center, on the inner edge of the Orion Arm, one of the spiral-shaped concentrations of gas and dust. The stars in the innermost 10,000 light years form a bulge and one or more bars that radiate from the bulge. The galactic centre is an intense radio source known as Sagittarius A*, a supermassive black hole of 4.100 (± 0.034) million solar masses. Stars and gases at a wide range of distances from the Galactic Center orbit at approximately 220 km per second. The constant rotation speed contradicts the laws of Keplerian dynamics and suggests that much (about 90%) of the mass of the Milky Way is invisible to telescopes, neither emitting nor absorbing electromagnetic radiation. This conjectural mass has been termed "dark matter". The rotational period is about 250 million years at the radius of the Sun. The Milky Way is moving at a velocity of approximately 600 km per second with respect to extragalactic frames of reference. The oldest stars in the Milky Way are nearly as old as the universe itself and thus probably formed shortly after the Dark Ages of the Big Bang.

The Sun is the star at the centre of the solar system. The Sun is 4.6 billion years old and the solar system with the planets is 4.5 billion years old. It is a nearly perfect sphere of hot plasma, heated to incandescence by nuclear fusion reactions in its core.

SOLAR SYSTEM AND PLANETS

Sun and Planets, with revolution time (Figure 2.15)

FIGURE 2.15   Proportional size, inclination & direction of rotation of planets.

KEPLER'S LAWS – Orbits of Planets

In astronomy, Kepler's laws of planetary motion, published by Johannes Kepler between 1609 and 1619, describe the orbits of planets around the Sun. The laws modified the much believed heliocentric theory of Nicolaus Copernicus, replacing its circular orbits and epicycles with elliptical trajectories, and explaining how planetary velocities vary. Kepler published his first two laws about planetary motion in 1609 in Astronomia nova and Kepler's third law was published in 1619. The three laws state that:

1. **All planets move in elliptical orbits with the Sun at one of the focal points.**
2. **The radius vector joining any planet to the Sun sweeps out in equal areas in equal times.**
3. **The square of the period of an orbit is proportional to the cube of the radius of that orbit.**

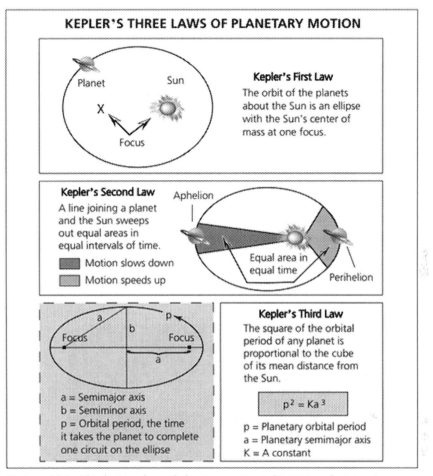

KEPLER'S THREE LAWS OF PLANETARY MOTION

Kepler's First Law
The orbit of the planets about the Sun is an ellipse with the Sun's center of mass at one focus.

Kepler's Second Law
A line joining a planet and the Sun sweeps out equal areas in equal intervals of time.

■ Motion slows down
▨ Motion speeds up

Equal area in equal time

Aphelion

Perihelion

Kepler's Third Law
The square of the orbital period of any planet is proportional to the cube of its mean distance from the Sun.

$$p^2 = Ka^3$$

p = Planetary orbital period
a = Planetary semimajor axis
K = A constant

a = Semimajor axis
b = Semiminor axis
p = Orbital period, the time it takes the planet to complete one circuit on the ellipse

Kepler published his first two laws of planetary motion in his 1609 work, *Astronima nova*. His third law appeared in his 1619 work, *Harmonice mundi*. The elliptical orbits shown here are exaggerated. A planet's true orbit is only slightly elliptic.

Knowledge of these laws, especially the second (the law of areas), proved crucial to Sir Isaac Newton in 1684–1685, when he formulated his famous law of gravitation between Earth and the Moon, and between the Sun and the planets, postulated by him to have validity for all objects anywhere in the universe. Newton showed that the motion of bodies subject to central gravitational force need not always follow the elliptical orbits specified by the first law of Kepler but can take paths defined by other, open conic curves; the motion can be in parabolic or hyperbolic orbits, depending on the total energy of the body. Thus, an object of sufficient energy – e.g., a comet – can enter the solar system and leave again without returning.

From Kepler's second law, it may be observed further that the angular momentum of any planet about an axis through the Sun and perpendicular to the orbital plane is also unchanging, distinguished them from his other discoveries. The usefulness of Kepler's laws extends to the motions of natural and artificial satellites, as well as to stellar systems and extrasolar planets. As formulated by Kepler, the laws do not consider the gravitational interactions (as perturbing effects) of the various planets on each other. The general problem of accurately predicting the motions of more than two bodies under their mutual attractions is quite complicated; analytical solutions of the three-body problem are unobtainable except for some special cases. It may be noted that Kepler's laws apply not only to gravitational but also to all other inverse-square-law forces and, if due allowance is made for relativistic and quantum effects, to the electromagnetic forces within the atom.

THE BLUE DOT – THE EARTH

> *It suddenly struck me that that tiny pea, pretty and blue was the Earth. I put up my thumb and shut one eye and my thumb blotted out the planet Earth. I did not feel like a giant. I felt very very small.* (Figure 2.16)

—Neil Armstrong

Five elements constitute the earth. Land, Water, Sky, Air and Fire. We call it *Pancha Boothas*. A human is also made of the five elements. Lewis Dartnell writes in his book 'Origins',

FIGURE 2.16 Earth seen from Moon.

The water in your body once flowed down the Nile, fell as monsoon rain onto India and swirled around the Pacific. The carbon in the organic molecules of your cells was mined from the atmosphere by the plants that we eat. The salt in your sweat and tears, the calcium of your bones and the iron in your blood all eroded out of the rocks of Earth's crust and the sulphur of the protein molecules in your hair and muscles was spewed out by volcanoes.

ORIGIN OF EARTH

As the universe was expanding, many galaxies and stars got formed. 4.6 billion years ago, the Sun, our star, was formed as a nearly perfect sphere of hot plasma, heated to incandescence by nuclear fusion reactions in its core, radiating the energy mainly as light and infrared radiation. It is by far the most important source of energy.

Earth formed 4.54 ± 0.07 billion years ago out of a flattened disc of dust orbited around the solar nebula, which just got formed. Earth formed when gravity pulled swirling gas and dust into what become the third planet from the Sun. Like its fellow terrestrial planets, Earth has a central core, a rocky mantle, and a solid crust.

The disc was intensely hot and volcanic, continuously bombarded by asteroids for few hundred million years. Over millions of years, the hot disc took shape of a sphere. Volcanic outgassing probably created the primordial atmosphere and then the ocean, but the early atmosphere contained almost no oxygen. Then, 4.527 billion years ago, a planet of size Mars, called **Theia** collided, and a piece came out as the Moon, and Earth's spin axis was tilted by 23.5°. (Figure 2.17)

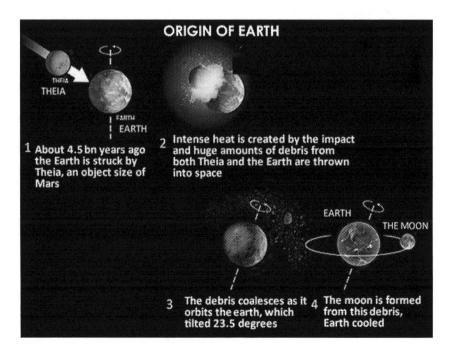

FIGURE 2.17 Origin of Earth.

Over the years the solar system was formed in an orderly manner with the planets revolving at constant speed in elliptical orbits. Several moons around planets also became spherical bodies. Earth-Moon formation as two spherical bodies with synchronised rotation evolved. About 4.4 billion years before, Earth cooled to the point, where liquid water could exist on its surface. Around 3.8 billion years, the Earth became bluish in colour, as it was fully covered with water. Microbial-induced sedimentary structures started appearing around 3.48 billion years before. They indicate the presence of a complex microbial ecosystem, most likely a purple layer of slime that thrived in the warm, wet, oxygen-free environment of early Earth, filling the atmosphere with sulphurous stench of anaerobic breath.

Carbon dating of the sedimentary rock confirmed the formation of early life started 3.7 billion years before. Most probably Mars also could have gone through this cycle. Inclinations in these two planets helped in varying temperature patterns over the surface.

The position, the synchronous rotation speed, and gravitation of the Moon are significant to help the process of evolution of life. Distance from the Earth to the Moon and the Sun, divided by the diameter of the Moon and Sun, respectively, **is 108**, making a perfect alignment of Sun-Earth-Moon system, as explained in Figure 2.18. Earth became a more stable planet. Step by step life moved to land, as land emerged at different parts of Earth, due to sudden movements of the Earth's tectonic plates. When the tectonic plates slide over one another, there is a cause of orogeny, which results in earthquakes and volcanoes. These disturbances cause vibrations, which spread in all the directions. As there is a relative motion of these plates, there is stress built up, which breaks by releasing the stored energy known as shock waves. These resulted land masses of varied types at different parts of earth.

The tilt is essential for four seasons (Figure 2.19) and for life. The tilt of Earth's axis changes over a 40,000-year interval. The shape of its orbit changes the Earth's distance from the Sun over a period of 100,000 years. Solar and lunar eclipses are shown in Figure 2.20.

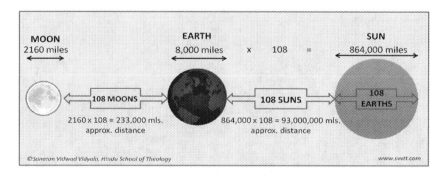

FIGURE 2.18   Significance of 108.

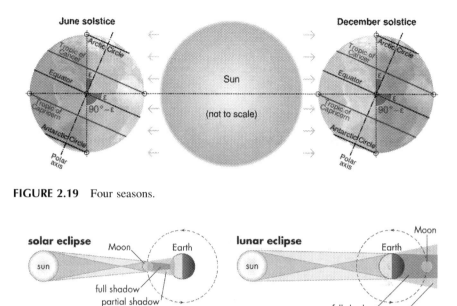

FIGURE 2.19   Four seasons.

FIGURE 2.20   Solar and lunar eclipses.

LIFE ON EARTH

Photosynthetic organisms appeared between 3.2 and 2.4 billion years ago and began enriching the atmosphere with oxygen. Life remained mostly small and microscopic until about 580 million years ago, when complex multicellular life arose, developed over time, and culminated in the Cambrian Explosion about 541 million years ago. This sudden diversification of life forms produced most of the major phyla known today and divided the Proterozoic Eon from the Cambrian Period of the Paleozoic Era. It is estimated that 99% of all species that ever lived on Earth have gone extinct. Mammals appeared 225 million years ago. Dinosaurs also appeared around this time. Estimates on the number of Earth's current species range from 10 million to 14 million, of which about 1.2 million are documented, but over 86% have not been described. However, it was recently claimed that 1 trillion species currently live on Earth, with only 1000th of 1% described. The timeline of life on Earth is depicted in Figure 2.21.

The Earth's crust has constantly changed since its formation, as has life since its first appearance. Species continue to evolve, taking on new forms, splitting into daughter species, or going extinct in the face of ever-changing physical environments. The process of plate tectonics continues to shape the Earth's continents and oceans and the life they harbour.

Life on Earth has suffered occasional mass extinctions at least since 542 million years ago. Although they were disasters at the time, mass extinctions have sometimes accelerated the evolution of life on Earth. When dominance of ecological niches

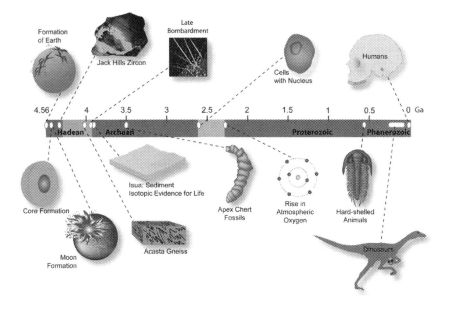

FIGURE 2.21   History of life on Earth.

Credit: Timeline courtesy of Andree Valley.

passes from one group of organisms to another, it is rarely because the new dominant group is "superior" to the old. It is usually because an extinction event eliminates the old dominant group and makes way for the new one.

The theory of evolution by natural selection, first formulated in Darwin's book "On the Origin of Species" in 1859, is the process by which organisms change over time because of changes in heritable physical or behavioural traits. Changes that allow an organism to better adapt to its environment will help it survive and have more offspring.

Evolution by natural selection is one of the best substantiated theories in the history of science, supported by evidence from a wide variety of scientific disciplines, including paleontology, geology, genetics, and developmental biology.

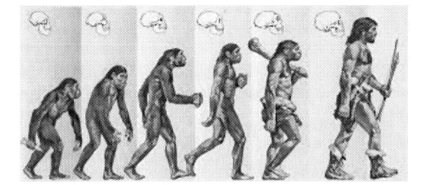

A small East African ape living around six million years ago was the last animal whose descendants would include both modern humans and their closest relatives, the chimpanzees. Only two branches of its family tree have surviving descendants. For reasons that are still unclear, apes in one branch developed the ability to walk upright. Brain size increased rapidly, and by two million years, the first animals classified in the genus Homo appeared. **From 250,000 years, anatomical humans are believed to have appeared — first in East Africa, a land connected with the Lemuria continent.**

FUTURE OF EARTH

Carl Sagan's *quote*

We live in a shooting gallery, surrounded by potential hazards. It is a matter of time, before large asteroids hit the Earth. If we are the only living on earth, what is the use of 99.99% of the universe. If our long-term survival is at stake, we have a basic responsibility to our species to venture to other worlds.

FIGURE 2.22 Earth is transforming.

Unlike all other life-forms on this planet, which must passively await their fate, we humans are masters of our own destiny. All living things will be extinct, including human beings. But humanity must survive.

*66 million years before a large Asteroid hit on Earth and the **dinosaurs** became extinct because they did not have a space program*

—Larry Niven

Mars was a beautiful planet with water, like Earth, harbouring life several million years before. **Life would have started on Mars before Earth.** See its fate today. Earth will also reach the stage of today's Mars in one million years or sooner.

Its fate in 2066 with the increased production of 51 billion tons of CO_2 per year can be seen in Figure 2.22. **Most men and women of this planet are trying their best to destroy Earth**. Earth and nature are already angry with humans. More natural calamities and new diseases will cause huge damage to life and the assets created by humans over the years. By the end of this century, many coastal cities will be flooded, as the sea water rises. Asteroids will pass by or hit the Earth. A large asteroid is on its way to take away one-third of Earth on 16 March 2880. We know how to deflect asteroids, but we cannot face the fury of nature. The Sun will not exist beyond four billion years, ending the solar system. While most of human race will get destroyed, a few will escape to a new planet of a different star, in the Milky Way galaxy, which will also collide with another galaxy in about five billion years.

EXOPLANETS

If we must protect humanity, we need to find an alternate Earth in another solar system, several light years away. Space exploration and space travel by humans is necessary to explore new frontiers. NASA is leading this venture, and several difficult possibilities are emerging. We will discuss this in the last chapter.

The basic requirement for exploration of the solar system and beyond is the ROCKET, the vehicle to send spacecraft, telescopes, and humans to travel to space. Let us learn about the rocket and the science behind it.

James Webb Space Telescope – a new era begins

DATA TO REMEMBER

Distance between the Sun and Earth = 1 AU = 1.496×10^8 km approx. 1.5 million km AU – Astronomical Unit
1 LY = 9.46 trillion km LY – Light Year = 63241 AU= 0.3066 parsecs.

Distance between the Sun and a nearby star = 1 Parsec = 3.26 LY

3 Brief History of Rockets

*What can be more fantastic than
to land on the surface of Mars.
A new great era will begin in
astronomy from the moment of
using rocket devices*

—Konstantin E. Tsiolkovsky (1896)

Father of Rocketry Konstantin E. Tsiolkovsky
Image Credit: K.E. Tsiolkovsky Museum, Kaluga, Russia

EARLY UNDERSTANDING

The word *rocket* could have come from *rocchetta*, a diminutive of Italian word *rocca* for distaff, meaning little spindle used to hold the thread from a spinning wheel. In the human evolution, men could not live without wars. Early wars were shows of human might. As things progressed, new tactics and new weapons came in. Rockets are the result of human ingenuity that have their roots in the early wars and scientific inventions of the past. They are natural outgrowths of necessity, knowledge, and many years of experimentation and research on rocket propulsion.

One of the first devices to successfully employ the principle of propulsion was aeolipile built by a Greek geometer and inventor, the Hero of Alexandria, in the first century. His device used steam as a propulsive gas. Hero mounted a hollow sphere on top of a water kettle, as shown in Figure 3.1. A fire below the kettle turned the water into steam, and the gas travelled through pipes to the sphere. Two L-shaped tubes on opposite sides of the sphere allowed the gas to escape, and in doing so gave a thrust to the sphere that caused it to rotate.

DOI: 10.1201/9781003323396-3

FIGURE 3.1   Hero engine.

CHINESE FIRE ARROWS

According to available records, the Chinese used fire arrows (Figure 3.2), using gun powder in the battle of Kaifeng-fu against Mongol invaders in 1232. The explosive used was a chemical mixture of potassium nitrate (saltpeter), sulphur, and charcoal, packed inside a bamboo tube attached to a long bamboo stick. Later, the use of gunpowder to create fire was attempted by other countries, in different forms, especially in canons.

FIGURE 3.2   Chinese fire arrows.

FIRST METAL ROCKETS FROM INDIA

As a real rocket with a metal chamber, filled with high-density compacted gunpowder, with an igniter and conical opening, it was used for the first time by Hyder Ali at Mysore, India in 1780. His son, Tipu Sultan, improved this rocket by attaching a sword as a warhead. This sophisticated rocket motor was 60 mm in

IRON CASE
- 2 KG GUN POWDER
- LENGTH: 250 MM
- DIAMETER: 60 MM,
- RANGE: 1.0-1.5 KM
- GUIDER SWORD
  BLADE (1m LONG)

FIGURE 3.3 Specifications of Indian rocket and the rocket at Woolwich royal artillery museum at London.

diameter and 250 mm long, with two kg of gunpowder. It had an igniter in the fore-end and a conical nozzle at the back-end. The payload was a one m-long metallic sword, tied to the rocket rigidly with a leather strap (Figure 3.3). The range of the rocket was more than one km. Three of the rockets were launched from a special metallic carriage with a guiding slot.

A large number of these rockets were produced and stored at multiple locations in Tipu's kingdom. He was the first to introduce Rocketeer Force with 5000 men (Figure 3.4). In the 1792 war against the British cavalries, at Srirangapatna in Mysore, more than 6000 rockets were fired as a barrage, and these uncontrolled rockets, making a terrific noise, pierced the horse cavalries at a distance of more than one km. Undoubtedly, the noise of rockets and fire frightened the horses and the British soldiers. The British men were defeated comfortably. When I went to Woolwich, I saw the original rocket and kept it with my chest with profound respect. The curator was kind enough to show the 1792 war details, written in golden letters, acknowledging the defeat of the British.

The reports of rockets in war in India boosted the European interest to rocketry in the 18th century and opened a new era of rocket development in the world. The success of Indian rocket barrages against the British in 1792 and again in 1799 caught the interest of artillery Colonel William Congreve. Nearly 10,000 rockets

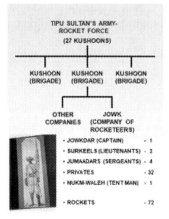

FIGURE 3.4 World's first rocket force – Mysore kingdom.

were shifted from Mysore kingdom to British Arsenal, after killing Tipu and destroying the explosive storage.

Congreve set out to reengineer Indian rockets in 1804 for use by the British military. **These modified Indian rockets were highly successful in battle.** They were used by the British Army against Napoleon in 1806 and later in ships to pound Fort McHenry during 1812–1814, against the Americans, which inspired Francis Scott Key to write "the rockets' red glare," in his poem that later became **The Star-Spangled Banner – the national anthem of the United States.**

Even with Congreve's work, the accuracy of rockets still had not improved much from the early days. The devastating nature of war rockets was not their accuracy or power, but their numbers, as used by Tipu. During a typical siege, thousands of them might be fired at the enemy. All over the world, Indian rockets drew attention of researchers to improve accuracy. An Englishman, William Hale, developed a technique called spin stabilisation. In this method, the escaping exhaust gases struck small vanes at the bottom of the rocket, causing it to spin much as a bullet does in flight.

Use of rockets continued to be successful in battles all over the European continent. Thus, the centre of rocket development shifted to Europe from India. Indian rocketry re-emerged only in the 1960s, in the free India.

BIRTH OF MODERN ROCKETRY

Many outstanding visionaries and scientific thinkers laid the foundations of rocket technology and space exploration in the world from the late 1980s.

PIONEER OF COSMONAUTICS: EQUATION

In 1896, a Russian school teacher, Konstantin Tsiolkovsky (1857–1935), proposed the idea of space exploration by rocket. In a report published in 1903, Tsiolkovsky suggested the use of liquid propellants for rockets in order to achieve greater range. Tsiolkovsky stated that the speed and range of a rocket were limited only by the exhaust velocity of escaping gases. For his ideas, careful research, and great vision, Tsiolkovsky has been called the Father of Modern Astronautics. He envisioned humans will land on Mars and find another Earth on which to live.

EARLY ROCKET DESIGNS

I went to Tsiolkovsky's house at Kaluga, 140 km from Moscow. Sitting on the old rotating chair used by him, I went through his earlier notes and saw the telescope used by him to see the stars, Mars, and the cosmic arena through the window. I was astonished to know his difficult life. I could realise the power of the mind of a disabled scientist to visualise man landing on Mars, even before aircraft flight. As a child, Konstantin caught scarlet fever and became hard of hearing. He was not accepted at elementary schools because of his hearing problem, so he was home schooled until the age of 16 by his mother. Nearly deaf, he worked as a high school mathematics teacher until retiring in 1920. Tsiolkovsky theorised many aspects of space travel and rocket propulsion in 1895. He was the first man to conceive **the space elevator**, after visiting Paris in 1895 and becoming inspired by the newly constructed Eiffel Tower.

TSIOLKOVSKY, THE FATHER OF ASTRONAUTICS

Tsiolkovsky on his own without the help of anyone figured out mathematics, physics, dynamics, and mechanics of a liquid engine rocket. He proposed the idea of space exploration reaching Mars. He calculated the exhaust velocity for escaping Earth using his unique design. His thoughts on space travel can be seen in his 1883 manuscript *Free Space*. His visionary article "Research into Interplanetary Space by means of Rocket Power" appeared in 1903 in the magazine *Nauchnoe Obozrenie* (Scientific Review). This brought him to the limelight.

TSIOLKOVSKY'S ROCKET EQUATION

Tsiolkovsky derived the so-called rocket equation, for calculating the exhaust velocity, based on the dynamics of the vehicle, the mass of which is decreasing in flight. It is this derivation that shows us the basis for rocket propulsion. He went on to describe multistage rockets, for velocity addition to attain the orbital velocity. The basic principle of a rocket – a device that can apply acceleration to itself using thrust by expelling part of its mass with high velocity – can thereby move due to the conservation of momentum (Figure 3.5).

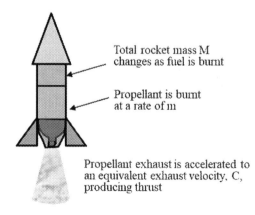

Total rocket mass M
changes as fuel is burnt

Propellant is burnt
at a rate of m

Propellant exhaust is accelerated to
an equivalent exhaust velocity, C,
producing thrust

FIGURE 3.5 Typical rocket system.

Tsiolkovsky's Rocket equation is

$$\Delta v = v_e \ln \frac{m_i}{m_f} = I_{sp} g_o \ln \frac{m_o}{m_f}$$

where:

Δv is the Maximum change of velocity (m/s)

m_i is the Initial total mass (including propellant) – "wet mass" (kg)

m_f is the effective exhaust velocity

$v_e = = I_{sp} g_o$ is the effective exhaust velocity (m/s); where

I_{sp} is the specific impulse

g_o is standard gravity

ln is the natural logarithm function.

Consider the rocket system shown in Figure 3.5. There is a rocket vehicle and propellant with total mass, *M in kg*, an effective exhaust velocity, *C in m/s*, and a mass flow rate in kg/s, *m (dot)*. The thrust F can be calculated in N, using Newton's Second Law as follows:

$$F_{(thrust)} = \dot{m} \cdot C = \frac{dM}{dt} C$$

$$F = Ma = M \frac{dv}{dt}$$

Therefore,

$$M \frac{dv}{dt} = -\frac{dM}{dt} C$$

So therefore,

$$dv = -C \frac{dM}{M}$$

As the rocket engine fires, the total mass of the rocket decreases, propellant is burnt, and the exhaust is accelerated.

Assuming that the rocket starts out with a velocity of v_o, ends with a velocity of v_f, has an initial mass of M_o, and has a final mass of M_f, then we can solve the above equation by integrating it through these limits:

$$ln(vo/vf)dv = -C \ ln(vo/vf)dM/M$$

Integrating and applying the limits, we get the rocket equation:

$$vf - vo = -C(ln \ Mf - ln \ Mo) = C \ ln(Mo/Mf)$$

Incremental velocity **Delta v = C ln (Mo/Mf)**

The basic equation for rocket propulsion, the Tsiolkovsky rocket equation, is named after him. This equation is used to estimate how much escape velocity is needed for interplanetary travels.

He calculated that the speed required to orbit the Earth is 7.9 km/s and that this could be achieved by means of a multi-stage rocket fuelled by liquid oxygen and liquid hydrogen. He realised at an early stage that a liquid-fuelled rocket with hydrogen and oxygen could produce considerable increase in exhaust velocity, needed for efficient space travel and escape from Earth. Among his works are designs for rockets with steering thrusters, multi-stage boosters, space stations, airlocks for exiting a spaceship into the vacuum of space, and closed-cycle biological systems to provide food and oxygen for space colonies.

He believed that colonising space would lead to the perfection of the human race, with immortality and a carefree existence. He said, "The Earth is the cradle of mankind, but one cannot eternally live in a cradle!"

"Man must at all costs overcome the Earth's gravity and have, in reserve, the space at least of the Solar System." This is great message for generations.

> Men are weak now, and yet they transform the Earth's surface. In millions of years, there might be increase to the extent that they will change the surface of the Earth, its oceans, the atmosphere and themselves. They will control the climate and the solar system just as they control the Earth. They will travel beyond the limits of our planetary system; they will reach other Suns and use their fresh energy instead of the energy of their dying luminary.

For his ideas, careful research, and great vision, Tsiolkovsky has been called the Father of Modern Astronautics. He inspired a generation of Soviet scientists, including Korolev.

> *The scientific legacy of Tsiolkovsky transferred to the Soviet State ... . is being creatively developed and successfully continued by Soviet Scientists*

—Sergei P Korolev 1957

His work influenced later rocketeers throughout Europe and America to work further.

ROBERT H GODDARD (1882–1945)

Early in the 20th century, an American, Robert H. Goddard conducted practical experiments in rocketry. He was interested to find a way of achieving higher altitudes than lighter-than-air balloons. He published a pamphlet in 1919 entitled "A Method of Reaching Extreme Altitudes". Today we call this mathematical analysis the meteorological sounding rocket. In his pamphlet, Goddard reached several conclusions important to rocketry. From his tests, he stated that a rocket operates with greater efficiency in a vacuum than in air. At the time, most people mistakenly believed that the presence of air was necessary for a rocket, just as they move on Earth.

Goddard believed that multistage or step rockets were the answer to achieving high altitudes and that the velocity needed to escape Earth's gravity could be achieved in this way. Goddard's earliest experiments were with solid-propellant rockets. In 1915, he began to try various types of solid fuels and to measure the exhaust velocities of the burning gases. While working on solid-propellant rockets, Goddard became convinced that a rocket could be propelled better by liquid fuel.

No one had ever built a successful liquid-propellant rocket before. It was a much more difficult task than building solid propellant rockets. Fuel and oxygen tanks, a feed system, and a combustion chamber would be needed. Despite the difficulties, Goddard achieved the first successful flight with a liquid propellant rocket on 16 March 1926. Fueled by liquid oxygen and gasoline, the rocket, shown in Figure 3.6, flew for only two and a half seconds, climbed 12.5 m, and landed 56 m away in a cabbage patch.

By today's standards, the flight was unimpressive, but like the first powered airplane flight by the Wright brothers in 1903, Goddard's gasoline rocket became the forerunner of a whole new era in rocket flight.

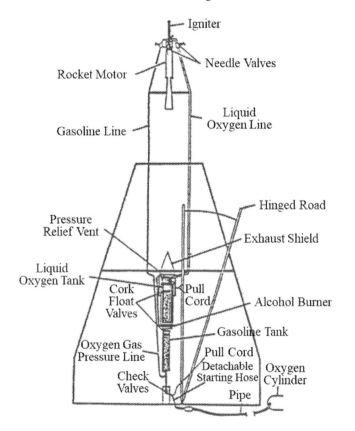

FIGURE 3.6   Dr. Goddard's 1926 rocket.

FIGURE 3.7   Robert Goddard with his liquid propellant-fueled rocket.

Goddard's experiments in liquid-propellant rockets continued for many years. His rockets grew bigger and flew higher (Figure 3.7). He developed a gyroscope system for flight control and a payload compartment for scientific instruments. Parachute recovery systems returned the rockets and instruments safely to the ground. We call Goddard the Father of Modern Rocketry for his achievements.

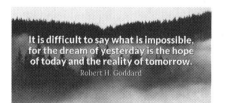

HERMANN OBERTH (1894–1989)

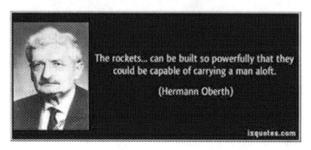

A third great space pioneer, Hermann Oberth, lived in Germany. He published a book in 1923 titled "The Rocket into Interplanetary Space," describing rocket dynamics to travel into outer space, with concept of the optimal velocity to escape Earth's gravity. He described the details of liquid-fuel rockets. Oberth introduced a number of ideas, such as rocket staging, film cooling of the engine walls, and strengthening the structure by pressurising propellant tanks. The book presented a detailed design of a two-stage rocket with all calculations, the first of its kind.

His effort became a reality in the latter half of the 20th century. Rockets were developed that were powerful enough to overcome the force of gravity to reach orbital velocities, paving the way for space exploration. Because of his writings, many small rocket societies sprang up around the world. In Germany, the formation of one such society, the Verein fur Raumschiffahrt (Society for Space Travel), in 1929, took up the development of the first guided missile of 360 km-range V-2. Oberth was the president of the society, and the team had Wernher von Braun and a team of rocket engineers.

In 1937, German engineers and scientists, including Oberth, assembled in Peenemunde on the shores of the Baltic Sea. There, under the directorship of Wernher von Braun, engineers and scientists built and flew the most advanced rocket of its time. The V-2 rocket (in Germany called the A-4, as shown in Figure 3.8) was the first guided missile. It achieved its great thrust by burning a mixture of liquid oxygen and alcohol + water at a rate of about one ton every seven seconds. A large number of them arched 100 km high over the English Channel at more than 5000 km per hour during World War II and attacked London.

After World War II, the United States and the Soviet Union created their own missile and space programs. In 1955, Oberth moved to Huntsville, USA, and joined von Braun in the US Army ballistic missile program. Oberth played an important

FIGURE 3.8   Oberth, von Braun and team and V2 – first guided rocket.

role in practical development of rocketry in Germany in the 1930s and inspired many European scientists.

SPACE ODYSSEY

As technologies advanced, our ability to investigate and unfold the space mysteries gradually increased. First, with telescopes, then with satellites, then space rovers, and ultimately humans in space. Men have set foot on the Moon, sent orbiting space stations, successfully landed rovers on Mars, and even photographed other galaxies. **How did it all happen?**

ORBITING SATELLITE

In the USSR, the rocket designer **Sergei Korolev** had developed the first-ever Inter Continental Ballistic Missile (ICBM), called the R7 (Figure 3.9). The space race between the Soviet Union and the USA began, giving an initial edge to the former.

This happened during the period of political hostility between the Soviet Union and the United States known as the Cold War. For several years, the two super-powers had been competing to develop different types of missiles, bombers, and nuclear submarines to carry nuclear weapons between continents.

This competition came to a head on with the launch of 83 kg satellite called Sputnik-1 by the USSR on 4 October 1957. Carried atop on R7 rocket, the Sputnik satellite sent out beeps from a radio transmitter, orbiting around Earth once every 96 minutes. Realising that the USSR had capabilities that exceeded US technologies, the Americans were worried. Then, a month later, on 3 November 1957, the Soviets achieved an even more impressive space venture. Sputnik II satellite, weighing 508 kg, carried a living creature, a dog Laika. It could not withstand 5 g acceleration and excess heating of the capsule and orbited around Earth but survived only for one day. The Sputnik-1 satellite and Laika are shown in Figure 3.10.

FIGURE 3.9   Sergei Korolev and his first R7 ICBM.

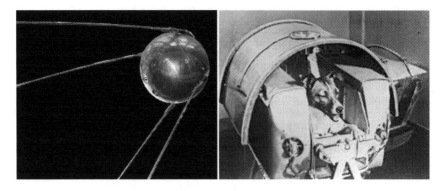

FIGURE 3.10   Sputnik 1–4 October 1957. Sputnik 2 with Laika – 3 November 1957.

The United States had been working on its own capability to launch a satellite, using a Navy-developed rocket called Vanguard. It was a dismal failure. After two failed attempts to launch a satellite into space, a satellite Explorer was orbited on 31 January 1958, by modified Jupiter C rocket, developed by a team consisted largely of German rocket engineers who had once developed ballistic missiles for Nazi Germany. Working for the U.S. Army at the Redstone Arsenal in Huntsville, Alabama, the German rocket engineers were led by Wernher von Braun and had developed the German V2 rocket into a more powerful rocket, called the Jupiter C, or Juno. Explorer-1 satellite carried several instruments into space for conducting science experiments (Figure 3.11).

FIGURE 3.11   Jupiter C with explorer-1.

One instrument was a Geiger counter for detecting cosmic rays. This was for an experiment operated by researcher James Van Allen, which, together with measurements from later satellites, proved the existence of what are now called the Van Allen radiation belts around Earth.

In 1958, space exploration activities in the United States were consolidated into a new government agency, the National Aeronautics and Space Administration (NASA). When it began operations in October 1958, NASA absorbed what had been called the National Advisory Committee for Aeronautics (NACA), and several other research and military facilities, including the Army Ballistic Missile Agency (the Redstone Arsenal) in Huntsville.

Yuri Gagarin

Russians continued to dominate the space program: the first artificial satellite, the first spacecraft on the Moon, the first docking of two satellites, the first man in space, and the first spacewalk, ahead of the USA.

The first human in space was the Soviet cosmonaut Yuri Gagarin, who made one orbit around Earth on 12 April 1961, on a mission that lasted 108 minutes and reached the height of 327 km. After one orbit, the command was given to decelerate as per the pre-set programme to land on land. The landing of Gagarin without any injury on land, gave the message that there was vast prospect for man's conquest of space. The USSR celebrated this feat, a clear lead over the USA.

The Soviet milestones included Luna 2, which became the first human-made object to hit the Moon in 1959. Soon after that, the U.S.S.R. launched Luna 3. Less than four months after Gagarin's flight in 1961, a second Soviet human mission orbited a cosmonaut around Earth for a full day. The USSR also achieved the first spacewalk and launched the Vostok 6 mission, which made **Valentina Tereshkova** the first woman to travel to space.

During the Soviet dominance, Vladimir N Chelomei was another leading chief designer of rockets, spacecraft, and space stations, and also ballistic and cruise missiles.

On 5 May 1961, NASA launched astronaut Alan Shepard into space, not on an orbital flight, but on a suborbital trajectory – a flight that goes into space but does not go all the way around Earth. Shepard's suborbital flight lasted just over 15 minutes

MAN ON MOON

Pained by the Soviet lead over the USA, on 25 May 1961, President John F. Kennedy challenged the Soviet Union to a Moon race, declaring an ambitious goal, "I believe that this nation should commit itself to achieving the goal, before the decade is out, of landing a man on the moon and returning him safely to Earth." What started with the Arms Race turned into the Space Race and now to the Moon Race.

Kennedy allotted $50 billion for manned mission to the Moon.

Wernher von Braun – The Architect of the Apollo Mission took up the responsibility to put man on the Moon and allow his safe return.

During the 1960s, NASA made progress toward President Kennedy's goal of landing a human on the Moon with a program called Project Gemini, in which astronauts tested technology needed for future flights to the Moon and tested their own ability to endure many days in spaceflight. Project Gemini was followed by Project Apollo, which took astronauts into orbit around the moon. In 1969, on Apollo 11, the United States sent the first astronauts to the Moon, and Neil Armstrong became the first human to set foot on its surface. Astronaut Neil Armstrong said "one giant leap for mankind" as he stepped onto the Moon. Six Apollo missions 11 to 17 (13 to return without landing) were made to explore the Moon between 1969 and 1972. During the landed missions, 12 astronauts collected samples of rocks and lunar dust for analysis.

The unfortunate death of visionary Chief Designer Korolev in January 1966 resulted in the end of the space-era lead for the Soviet Union of outstanding successes balanced by tremendous risk. Voskhod launcher planned to be ready for a manned mission to the Moon before the USA, but it could not be realised. This put the USA ahead of the USSR.

During the 1960s and 1970s, NASA also launched a series of space probes called Mariner, which studied Venus, Mars, and Mercury. During the 1970s, NASA also carried out Project Viking in which two probes landed on Mars, took numerous photographs, examined the chemistry of the Martian surface environment, and tested the Martian dirt (called regolith) for the presence of microorganisms.

SPACE SHUTTLE

America took another lead over the USSR in establishing partial reusability in manned missions and launching of satellites, interplanetary probes, and the Hubble Space Telescope (HST), through the Space Shuttle program, carrying out 135 missions during the period 1981–2011. Of the five orbiters built, two were lost in mission accidents: the Challenger in 1986 and the Columbia in 2003, with a total of 14 astronauts killed. The Space Shuttle was retired from service upon the conclusion of Atlantis's final flight on 21 July 2011. The U.S. relied on the Russian Soyuz spacecraft to transport astronauts to the ISS until the launch of the Crew Dragon Demo-2 mission in May 2020 on a SpaceX Falcon 9 rocket as part of the Commercial Crew Program.

SPACE STATIONS

Space stations marked the next phase of space exploration. The Soviets launched the first modular three-crewed space station **MIR**, assembled in orbit in 1986, and it was in orbit till 2001. Over its lifetime, the space station hosted 125 cosmonauts and astronauts from 12 different nations, including India. It supported 17 space expeditions, including 28 long-term crews. This was followed by NASA's **Skylab** space station, the first orbital laboratory in which astronauts and scientists studied Earth and the effects of spaceflight on the human body.

Human space exploration is presently limited to low-Earth orbit, where many countries participate and conduct research on the International Space Station (ISS), a modular space station in 400 km Earth orbit, from November 1998. It is a multinational collaborative project involving five participating space agencies: NASA (USA), Ruscosmos (Russia), JAXA (Japan), ESA (Europe), and CSA (Canada). The ownership and use of the space station were established by intergovernmental treaties and agreements, the main mission control at Moscow. The ISS serves as a microgravity and space environment research laboratory in which scientific research is conducted in astrobiology, astronomy, meteorology, physics, and other fields. As of November 2020, 242 astronauts, cosmonauts, and space tourists from 19 different nations have visited the space station, many of them multiple times; this includes 152 Americans, 49 Russians, nine Japanese, eight Canadians, and five Italians. At a time, there will be seven astronauts in the ISS.

CHINA'S TIANGONG SPACE STATION

Tiangong is a space station placed in low Earth orbit at 425 km above the surface. The Tiangong Space Station, launched on 29 April 2021, has three astronauts in orbit making the station operational. Once completed, it will be roughly one-fifth the mass of the International Space Station and about the size of the decommissioned Russian Mir space station. It will be fully operational in 2022 with three to six crew members.

SATELLITES IN BRIEF (DETAILS IN SEPARATE CHAPTER)

Earth and the Moon are examples of natural satellites. Thousands of artificial, or man-made, satellites orbit Earth. They provide remote Earth observation, communication, and navigation support. Some take pictures of the planet that help meteorologists predict weather and track hurricanes. Some take pictures of other planets, the Sun, or faraway galaxies. Astronomical satellites found new stars and gave a new view of the centre of the galaxy. These pictures help scientists better understand the solar system and universe.

EARTH OBSERVATION SATELLITES

Satellites could see large areas of Earth at one time. This ability means satellites can collect more data, more quickly, than instruments on the ground. This helps in resources mapping for development and better utilisation. Satellites discovered an ozone hole over Antarctica, pinpointed forest fires, and gave photographs of the nuclear power plant disaster at Chernobyl in 1986.

COMMUNICATION, NAVIGATION

Other types of satellites are used for communications, by beaming visual signals and voice-phone calls around the world. A group of more than 20 satellites make up the Global Positioning System, or GPS. If you have a GPS receiver, these satellites can help figure out your exact location.

In the 1980s, satellite communications expanded to carry television programs, and people were able to pick up the satellite signals on their home dish antennae.

The Gulf War proved the value of satellites in modern conflicts. During this war, allied forces were able to use their control of the "high ground" of space to achieve a decisive advantage. Satellites were used to provide information on enemy troop formations and movements, early warning of enemy missile attacks, and precise navigation in the featureless desert terrain. The advantages of satellites allowed the coalition forces to quickly bring the war to a conclusion, saving many lives.

Space systems continue to become more and more integral to homeland defence, weather surveillance, communication, navigation, imaging, and remote sensing of disasters. As the demand for more and larger payloads increase, a wide array of powerful and versatile rockets is needed. The satellite launch service becomes a component of economy in the competitive world.

Satellite Launch Vehicles

Launch vehicles (LV) are used to carry spacecraft from the surface of the Earth into space. To achieve a viable orbit, it is necessary to inject the spacecraft with a velocity of about 8 km/s. The launch vehicle either puts a spacecraft into its final trajectory or into a parking or intermediate trajectory or orbit from which an on-board propulsion transfer system is used to achieve the final trajectory.

ISRO

SLV-3 India's first satellite LV orbited 40 kg Rohini satellite in Earth's orbit on 18 July 1980 from Sriharikota range, establishing indigenous launch capability and making India the seventh member of the space club, after the Soviet Union (Oct 1957), USA (Feb 1958), France (Nov 1965), Japan (Feb 1970), China (Apr 1970), and the UK (Oct 1971). Now, India is one of the leading four space-faring nations with reliable operational PSLV, GSLV Mk II, and GSLV Mk III. So far, there were 79 launch missions from India's space port, named after Satish Dhawan, 119 Indian spacecraft, 342 foreign satellites from 35 countries, 13 student satellites, and two re-entry missions, using these launchers. Development is in progress for a high-thrust semi-cryogenic engine to be used as the core stage with two solid propellant boosters to upgrade payload delivery capability. A Small Satellite Launch Vehicle (SSLV) is also getting ready to tap small satellites (up to 500 kg) market potential. Operational ISRO launchers as of now and proposed ones are shown in Figures 3.12 and 3.13.

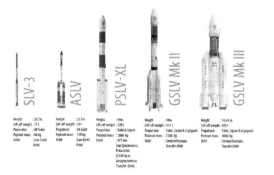

FIGURE 3.12   ISRO launchers.

Courtesy: ISRO.

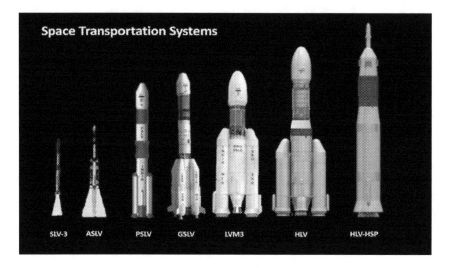

FIGURE 3.13   ISRO's newer HLV – in development beyond GSLV Mk-III.

Since the earliest days of discovery and experimentation, rockets have evolved from simple gunpowder devices into giant vehicles capable of travelling into outer space. **Rockets have opened the universe to direct exploration by humankind, as envisioned by Tsiolkovsky.**

LAUNCH VEHICLES OF THE WORLD

Many space-faring countries are in competition to market their launch capabilities of rockets for satellites to place in orbit. Scientific experiments have also started for exploring the planets of the solar system to find life, and beyond to find another Earth for human settlement. The world scenario of rockets in different countries is shown in Figure 3.14.

Private rockets for satellites, cargo, manned missions, and space exploration have emerged in recent times.

With the Artemis program, as shown in Figure 3.15, NASA will land men and the first woman on the Moon by 2024–2025, using innovative technologies to explore more of the lunar surface than ever before. Then, NASA will take the next giant leap – sending astronauts to Mars.

As part of the Artemis program, NASA is planning to send its first mobile robot to the Moon in late 2023 in search of ice and other resources on and below the lunar surface at the south pole. Data from the Volatiles Investigating Polar Exploration Rover (VIPER) will help the agency map resources at the lunar South Pole that could one day be harvested for long-term human exploration at the Moon.

VIPER's design calls for using the first headlights on a lunar rover to aid in exploring the permanently shadowed regions of the Moon. These areas haven't seen sunlight in billions of years and are some of the coldest spots in the solar system.

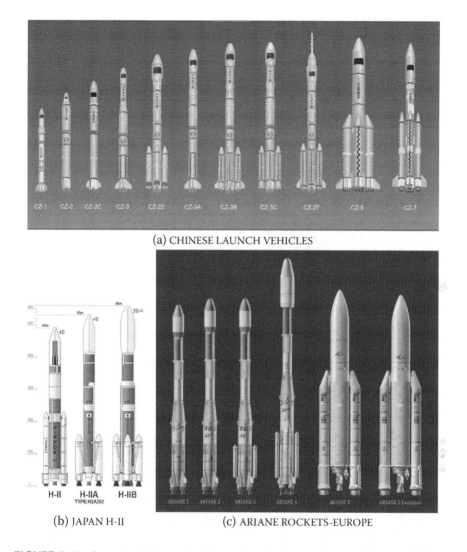

(a) CHINESE LAUNCH VEHICLES

(b) JAPAN H-II

(c) ARIANE ROCKETS-EUROPE

FIGURE 3.14 Launch vehicles – world scenario. (a) Chinese Launch Vehicles. (b) Japan H-II. (c) Ariane Rockets – Europe.

Image Source: wikimedia.org

Chinese Launch Vehicles Image Source: space.com

Running on solar power, VIPER will need to quickly manoeuvre around the extreme swings in light and dark at the lunar South Pole.

NASA's backbone for deep space exploration is the biggest rocket ever built, the Space Launch System (SLS), as shown in Figure 3.16, with the Orion spacecraft and the Gateway lunar command module.

In the half-century since people visited the Moon, NASA has continued to push the boundaries of knowledge to deliver on the promise of American ingenuity and

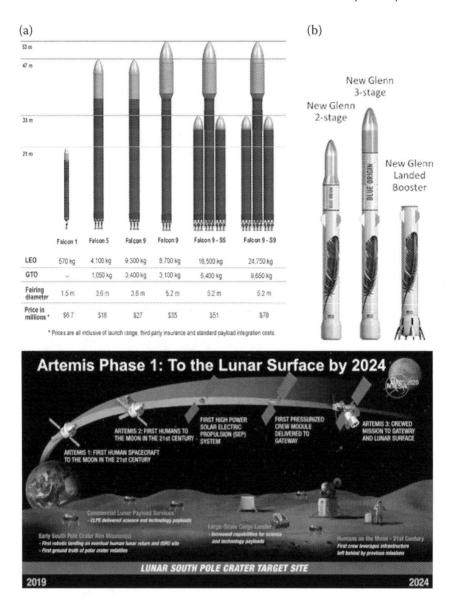

FIGURE 3.15   Launch vehicles – USA and **NASA PLAN – MOON 2024 – Artemis**.

Sourse: NASA a) Space X Falcon (USA) Image Credit: www.spacex.com b) Blue Origin (USA) Image Credit: space.nss.org.

leadership in space. And NASA will continue that work by moving forward to the Moon with astronauts landing on the lunar South Pole by 2024.

NASA is implementing the President's Space Policy Directive-1 to "lead an innovative and sustainable program of exploration with commercial and international partners to enable human expansion across the solar system."

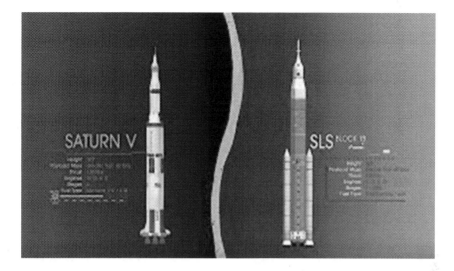

FIGURE 3.16   New rocket (SLS) for moon mission – 2024.

Source: NASA.

With its partners, NASA will use the Gateway lunar command module orbiting the Moon as a staging point for missions that allow astronauts to explore more parts of the lunar surface than ever. This work will bring new knowledge and opportunities and inspire the next generation. The Moon will provide a proving ground to test technologies and resources that will take humans to Mars and beyond, including building a sustainable, reusable architecture.

WORLD ROCKETS AT A GLANCE

Rockets from the USA, Russia, China, Japan, India, and Europe through their evolution are shown in Figure 3.17: Other countries with satellite launch capability are Israel, Iran, and North Korea.

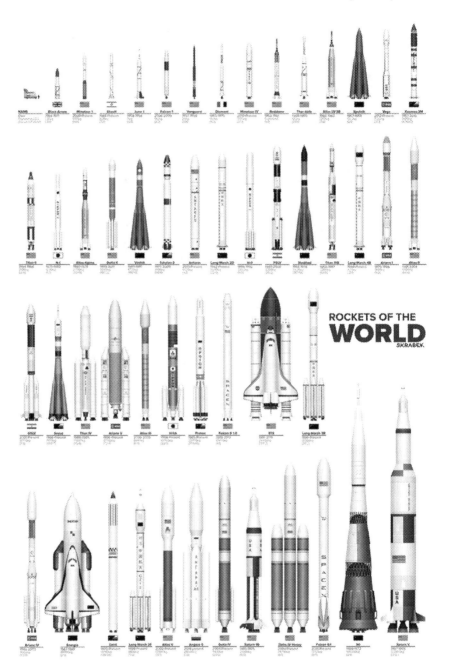

FIGURE 3.17   Rockets of the world.

Image Credit: Tyler Skrabek.

4 Rocket Principles

WHAT IS A ROCKET? BASIC ROCKET SCIENCE

In Indian villages, mothers used to feed their children, showing them the Moon and stars. The excited children used to call the Moon their uncle, who shows himself with varying sizes in a month. Today's children look at their smart phones and eat. Did we ever think it was possible for humans to visit the Moon, until 20 July 1969, when Neil Armstrong and Buzz Aldrin landed on the Moon and safely returned? The 20th century brought forth aircraft, computers, nuclear weapons, and also rockets that have placed humans in a different league on Earth.

We understand now that rockets can help us to explore the universe, the Sun, the Moon, and planets and finally escape the bonds of Planet Earth, our cradle, and go into the new frontiers that have mystified mankind for so long. Rockets also help put satellites in orbit around the Earth for communication, navigation, Earth resource mapping, and disaster warning. No doubt that these rockets are real assets.

We often hear in recent times that India launched a rocket and placed a satellite in an orbit above Earth. The questions come to mind: what is a satellite, and what is a rocket? Satellite or spacecraft is a system that goes around the Earth in an orbit or goes beyond the Earth on space exploration. The very purpose of a rocket is to impart the required velocity to the satellite to reach its orbit, as specified in the mission (Figure 4.1).

Let us learn about a rocket. The rocket is a flying vehicle powered by a self-contained device propelled by a reaction engine. The engine in its simplest form is a chamber enclosing a gas under pressure and an opening at one end, called a nozzle. When the pressurised gas is allowed to escape through the nozzle, the rocket is pushed in the opposite direction.

PRINCIPLE

An inflated balloon provides a simple way to understand the concept of a rocket. When closed completely, the air inside the balloon is compressed by the balloon's

DOI: 10.1201/9781003323396-4

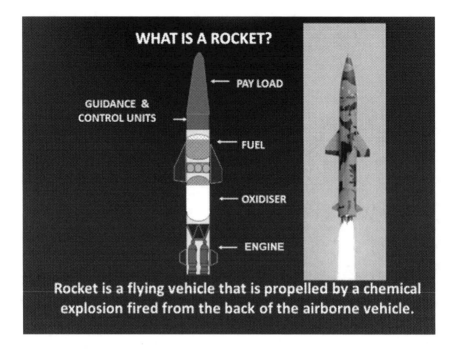

FIGURE 4.1   Parts of a typical rocket.

rubber walls. The air pushes back so that the inward and outward pressing forces are balanced. This balanced condition is called a state of mechanical equilibrium. When the pressurised air is suddenly released through the narrow opening (nozzle), air escapes through it into the surrounding lower pressure air, and the balloon is propelled in the opposite direction due to the reaction force, called thrust.

We can see from Figure 4.2 that action leads to reaction. The functioning of the rocket is similar except the way the pressurised gas is produced. With a space

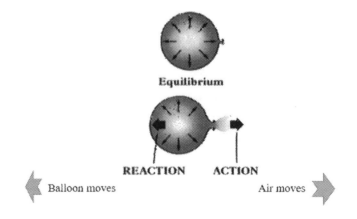

FIGURE 4.2   Balloon experiment.

rocket, the gas is produced by burning a chemical substance called propellants in the combustion chamber, which is released rearward through a nozzle.

NEWTON'S LAWS OF MOTION

The science of rocketry was explained when the great English scientist Sir Isaac Newton published a book entitled *Philosophiae Naturalis Principia Mathematica* in 1687. Newton stated three important scientific principles that govern the motion of all objects, whether on Earth or in space. These principles are called Newton's Laws of Motion. The function of a rocket is based on these laws. The Laws are:

1. **Objects at rest will stay at rest and objects in motion will stay in motion in a straight line unless acted upon by an unbalanced force.**
2. **Force is equal to mass times acceleration.**
3. **For every action there is always an opposite and equal reaction.**

So, the rocket is a vehicle, typically cylindrical, containing propellants that produce hot gases that are ejected rearward through a nozzle and, in doing so, create an action force accompanied by an opposite and equal reaction force pushing the vehicle forward. The rocket is able to operate in outer space because the propellant inside has both fuel and oxidizer. Let us discuss the laws in more detail.

NEWTON'S FIRST LAW – "THE LAW OF INERTIA"

An object in a state of uniform motion tends to remain in that state of motion unless an external unbalanced force is applied to it. Likewise, an object at rest will remain at rest unless an outside force is acting on it. It is necessary to understand the terms rest, motion, and external unbalanced force.

Rest is the state of an object when it is not changing its position in relation to its surroundings. If you are sitting in a chair, you can be said to be at rest. This term, however, is relative. Your chair may actually be one of many seats on a speeding airplane. The important thing to remember here is that you are not moving in relation to your immediate surroundings. Let us ignore Earth's rotation.

Motion: Motion is defined as an object changing position in relation to its surroundings. A ball is at rest if it is just lying on the ground. The ball is in motion if it is rolling. In Figure 4.3, a rolling ball changes its position in relation to its immediate surroundings. When a rocket blasts off the launch pad, it changes from a state of rest to a state of motion.

The third term in the law is external unbalanced force. If you hold a ball in your hand and keep it still, the ball is at rest. All the time the ball is held there though, it is being acted upon by forces. The force of gravity is trying to pull the ball downward, while at the same time your hand is pushing against the ball to hold it up. The forces acting on the ball are balanced. Let the ball go, or move your hand upward, and the forces become unbalanced. The ball then changes from a state of rest to a state of motion.

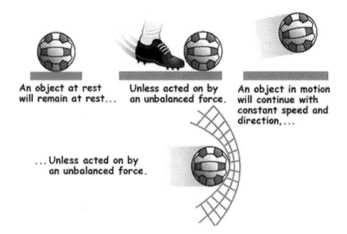

FIGURE 4.3 Newton's first law of motion.

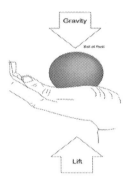

Let us consider a rocket. The rocket on the launch pad is balanced, as the surface of the pad pushes the rocket up while gravity tries to pull it down. As the engines are ignited, the thrust from the rocket unbalances the forces, and the rocket travels upward.

NEWTON'S SECOND LAW – "F = MA"

The rate of change of momentum is proportional to the force impressed upon an object and is in the same direction of the force.

This law of motion is essentially a statement of a mathematical equation. The three parts of the equation are mass (m), acceleration (a), and force (F). Using letters to symbolise each part, the equation can be written as follows:

$$F = ma. \quad (F \text{ and } `a' \text{ are vectors})$$

By using simple algebra, we can also write the equation two other ways:

$$a = F/m, \text{ also } m = F/a$$

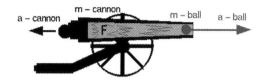

FIGURE 4.4   Forces of a cannon.

To explain this law, we will use the common example of a cannon, as shown in Figure 4.4.

When the cannon is fired, an explosion propels a cannon ball out of the open end of the barrel. It flies a kilometer or two to its target. At the same time, the cannon itself is pushed backward a meter or two. This is action and reaction at work (third law). The force acting on the cannon and the ball is the same. What happens to the cannon and the ball is determined by the second law. Look at the two equations below.

$$f = m(cannon) * a(cannon)$$

$$f = m(ball) * a(ball)$$

The first equation refers to the cannon and the second to the cannon ball.

In the first equation, the mass is the cannon itself and the acceleration is the movement of the cannon. In the second equation, the mass is the cannon ball, and the acceleration is its movement. Because the force (exploding gun powder) is the same for the two equations, the equations can be combined and rewritten below.

$$m(cannon) * a(cannon) = m(ball) * a(ball)$$

In order to keep the two sides of the equations equal, the accelerations vary with mass. In other words, the cannon has a large mass and a small acceleration. The cannon ball has a small mass and a large acceleration.

Let's apply this principle to a rocket. Replace the mass of the cannon ball with the mass of the gases being ejected out of the rocket engine. Replace the mass of the cannon with the mass of the rocket moving in the other direction. Force is the pressure created by the controlled combustion taking place inside the rocket's engine. That pressure accelerates the gas one way and the rocket the other.

Newton's second law of motion is especially important to design efficient rockets. The exit velocity is a measure of engine performance to enable a rocket to climb into low Earth orbit, by achieving a speed in excess of 28,000 km per hour. A speed of over 40,250 km per hour is necessary to enable a rocket to escape from Earth and travel out into deep space. Attaining space flight speeds requires the rocket engine to achieve the greatest action force possible in the shortest time. In other words, the engine must burn a large mass of fuel and push the resulting gas out of the engine as rapidly as possible.

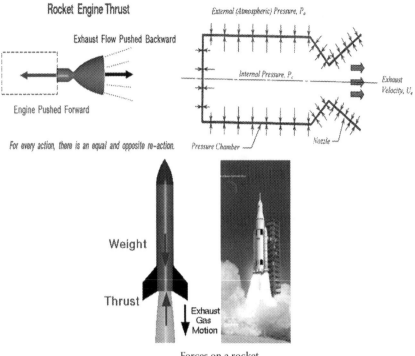

Forces on a rocket

FIGURE 4.5   Newton's third law of motion.

Image Credit: grc.nasa.gov.

NEWTON'S THIRD LAW – "THE LAW OF ACTION AND REACTION"

This law states that every action has an equal and opposite reaction. If you have ever stepped off a small boat that has not been properly tied to a pier, you will know exactly what this law means. The law is explained by thrust of an engine and rocket opposite to exhaust in Figure 4.5.

A rocket can lift off from a launch pad only when it expels gas out of its engine. The rocket pushes on the gas (action), and the gas in turn pushes on the rocket (reaction). With rockets, the action is the expelling of gas out of the engine, resulting in force (F). The reaction is the movement of the rocket in the opposite direction. To enable a rocket to lift off from the launch pad is the thrust from the engine, which must be greater than the mass of the rocket. Thrust (T) divided by weight (W) is an important factor for rocket take off, and it must be more than 1 (T/W > 1). In space, however, even tiny thrusts will cause the rocket to change direction.

FORCES ON A ROCKET DURING INITIAL FLIGHT

When a rocket is in flight in the atmosphere, four forces act on it: weight, thrust, and the two aerodynamic forces, lift and drag (Figure 4.6).

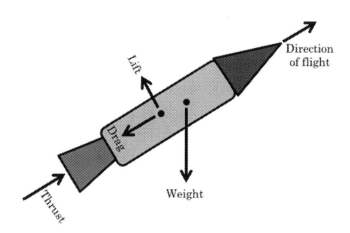

FIGURE 4.6   Force acting on a rocket.

The amount of the weight depends on the mass of all of the parts of the rocket. Thrust generated by propulsion system works the opposite of weight. Aerodynamics explains the motion of air and the forces on bodies moving through the air. Lift is the aerodynamic force that works in a 90° angle to the direction of the flight. Lift is not as significant a force on a rocket as it is on an airplane, which has large wings and wider body. Drag is the aerodynamic force that works against the upward movement of the rocket, and it largely depends on its configuration design. Rockets actually work better in space than they do in air, due to the absence of aerodynamic forces. In space, the exhaust gases can escape freely.

CENTRE OF GRAVITY AND CENTRE OF PRESSURE

In addition to centre of gravity (CG), there is another important centre inside the rocket that affects its flight. This is the centre of pressure (CP). The centre of pressure exists only when air is flowing past the moving rocket. This flowing air, rubbing and pushing against the outer surface of the rocket, can cause it to begin moving around one of its three axes.

Let us consider a weathervane, which is an arrow-like stick that is mounted on a rooftop and used for showing wind direction. The arrow is attached to a vertical rod that acts as a pivot point. The arrow is balanced so that the centre of mass is right at the pivot point. When the wind blows, the head of the arrow points towards the on-coming wind. The tail of the arrow points in the downwind direction as shown below.

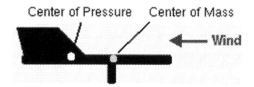

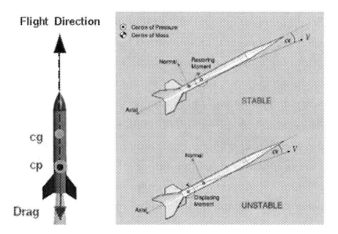

FIGURE 4.7 Positions of centre of gravity and centre of pressure.

Image Credit: NASA.

The reason that the weathervane arrow points into the wind is that the tail of the arrow has a much larger surface area than the arrowhead. The flowing air imparts a greater force to the tail than the head, and therefore the tail is pushed away. There is a point on the arrow where the surface area is the same on one side as the other. This spot is called the centre of pressure (CP). The centre of pressure is not in the same place as the centre of mass. If it were, then neither end of the arrow would be favoured by the wind, nor would the arrow not point. The centre of pressure is between the centre of mass and the tail end of the arrow, which has more surface area than the head end.

In a rocket, it is extremely important that the centre of pressure be located toward the tail and the centre of mass be located toward the nose, as shown in Figure 4.7. The rocket will be stable in flight. If they are in the same place or very near each other, then the rocket will be unstable in flight. The rocket will then try to rotate about the centre of mass in the pitch and yaw axes, leading to a dangerous situation. With the centre of pressure located in the right place, the rocket will remain stable.

Larger rockets have control systems to help them steer in the desired trajectory and remain stable.

THE ROCKET EQUATION

THRUST OF ROCKET

Thrust is nothing but the force produced by the propulsion system acting on the rocket. The unit of thrust is Newtons. This is the force generated to overcome other forces like gravity, drag in order to move forward in the specified trajectory for achieving its mission. In other words, thrust is the net force acting on the rocket, calculated as the rate of change of momentum, as per Newton's second law.

Momentum is a vector quantity and is defined as the product of mass × velocity. p = mv (p is momentum of mass m, moving with scalar velocity v)

$$F = \frac{dp}{dt} = m\frac{dv}{dt} + v\frac{dm}{dt}$$

(based on conservation of momentum with change in velocity and change in mass, as in a rocket flight).

F = m(dot)v(exit) + mv(exit dot).

The thrust for the rocket continues as long as its engines are firing. Furthermore, the mass of the rocket changes during flight, as the propellant is consumed and the exit velocity increases. Its mass is the sum of all its parts. Rocket parts include: engines, propellant tanks, payload, control system, and propellants. By far, the largest part of the rocket's mass is its propellants. But that amount constantly changes as the engine fires. That means that the rocket's mass gets smaller during flight. The acceleration of the rocket has to increase as its mass decreases. That is why a rocket starts off moving slowly and goes faster and faster as it climbs into space.

The first term m (dot) v (exit) is the momentum thrust represented by the product of the propellant mass flow rate and its exhaust velocity relative to the vehicle. This force represents the total propulsion force when the nozzle exit pressure equals the ambient pressure.

Because of fixed nozzle geometry and changes in ambient pressure due to variations in altitude, there can be an imbalance of the external environment or atmospheric pressure p (out) and the local pressure p (exit) of the hot gas jet at the exit plane of the nozzle. So, the second term represents the pressure thrust consisting of the product of the cross-sectional area at the nozzle exit (where the exhaust jet leaves the vehicle) and the difference between the exhaust gas pressure at the exit and the ambient fluid pressure. v (exit dot) is the acceleration component with mass of gas exiting at the exhaust of the divergent nozzle. This additional force can also be written as pressure multiplied by area. That is (p (exit) – p (out)) A (exit), if exit pressure (p exit) is more than outside pressure (p out). A (exit) is exit cross sectional area of the divergent nozzle. If the exhaust pressure is less than the surrounding fluid pressure, the pressure thrust is negative. Because this condition gives a low thrust and is undesirable, the rocket nozzle is usually so designed that the exhaust pressure is equal or slightly higher than the ambient fluid pressure.

Figure 4.8 describes the thrust equation.

This is the most important equation for rocket design.

When the ambient atmosphere pressure is equal to the exhaust pressure, the pressure term is zero and the thrust is

$$F = m(dot)v(exit).$$

In the vacuum of space p (out) = 0 and the thrust becomes

$$F = m(dot)v(exit) + p(exit)A(exit)$$

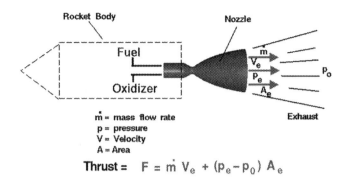

FIGURE 4.8   Thrust equation.

Image Credit: grc.nasa.gov.

ACCELERATION

A rocket's acceleration depends on three major factors, consistent with the equation for acceleration of a rocket. First, the greater the exhaust velocity of the gas relative to the rocket, v (exit), the greater the acceleration is. The practical limit for v (exit) is about 2.5×10^3 m/s for conventional (non-nuclear) hot-gas propulsion systems. The second factor is the thrust-rate at which mass is ejected from the rocket. The faster the rocket burns its fuel, the greater its thrust, and the greater its acceleration. The third factor is the mass m of the rocket. The smaller the mass is (all other factors being the same), the greater the acceleration. The rocket mass m decreases dramatically during flight because most of the rocket is fuel to begin with, so that acceleration increases continuously, reaching a maximum just before the fuel is exhausted (Figure 4.9).

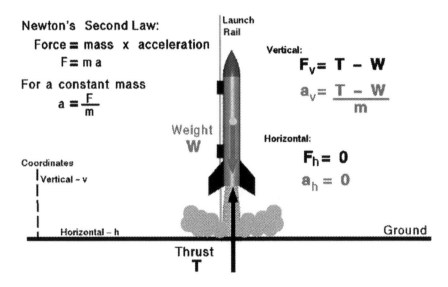

FIGURE 4.9   Acceleration at lift off.

Image Credit: grc.nasa.gov.

The rocket has a mass m and an upward velocity v. The net external force on the system is −mg, if air resistance is neglected. A time Δt later the system has two main parts, the ejected gas and the remainder of the rocket. The reaction force on the rocket is what overcomes the gravitational force and accelerates it upward.

By calculating the change in momentum for the entire system over Δt, and equating this change to the impulse, the following expression can be shown to be a good approximation for the acceleration of the rocket.

$$a = (v(exit)m(dot))/m - g$$

where **a** is the acceleration of the rocket, v(exit) is the escape velocity, m is the mass of the rocket, m (dot) is the rate of change of mass of the ejected gas.

TOTAL IMPULSE

It is the thrust force F (which can vary with time) integrated over the burning time t.

$$I = F (Integral\ 0 - t)\ dt)$$

For constant thrust and negligible start and stop transients this reduces to

$$I = Ft$$

It is proportional to the total energy released by all the propellant in a propulsion system.

SPECIFIC IMPULSE

Specific impulse is a measure of how efficiently the propellant creates thrust. It is the thrust produced per unit rate of consumption of the propellant that is usually expressed in kg-s of thrust per kg of propellant and that is a measure of the efficiency of a rocket engine.

Rocket Thrust Equation

Rocket Thrust Equation $\quad F = \dot{m} V_e + (p_e - p_a) A_e$

where p = pressure, V = velocity, A = area, $\dot{m}$ = mass flow rate, F = thrust

Define: Equivalent Velocity: $\quad V_{eq} = V_e + \dfrac{(p_e - p_a) A_e}{\dot{m}} \qquad F = \dot{m} V_{eq}$

Define: Total Impulse: $\quad I = F \backslash t = \displaystyle\int F\,dt = \int \dot{m} V_{eq}\,dt = m V_{eq}$

Define: Specific Impulse: $\dfrac{\text{Total Impulse}}{\text{Weight}} \qquad Isp = \dfrac{I}{m g_a} = \dfrac{V_{eq}}{g_a} \qquad$ units = sec

$$\boxed{Isp = \dfrac{F}{\dot{m} g_a}}$$

where:

F is the thrust obtained from the engine (newtons)

g_o is the standard gravity, which is nominally the gravity at Earth's surface (m/s2),

Isp is the specific impulse measured (seconds),

m (dot) is the mass flow rate of the expended propellant (kg/s or slugs/s)

Solid and liquid propellants have limited Isp of the order of 200–300 seconds and cryogenic propellants LH2+LOX has 450–460 seconds (in vacuum). An example of a specific impulse measured in time is 453 seconds, which is equivalent to an effective exhaust velocity of 4440 m/s, for the RS-25 engines when operating in a vacuum. https://en.wikipedia.org/wiki/Specific_impulse-cite_note-33 An air-breathing jet engine typically has a much larger specific impulse of the order of 6000 seconds or more at sea level.

An air-breathing engine is thus much more propellant efficient than a rocket engine, because the free air serves as reaction mass and oxidizer for combustion. While the *actual* exhaust velocity is lower for air-breathing engines, the effective exhaust velocity is very high for jet engines. This is because the effective exhaust velocity calculation assumes that the carried propellant is providing all the reaction mass and all the thrust. The ramjet and scramjet air breathing engines have comparatively less specific impulse, but can operate at higher Mach number.

It has been reported that the highest specific impulse for a chemical propellant ever test-fired in a rocket engine was 542 seconds (5.32 km/s) with a rare tri-propellant of lithium, fluorine, and hydrogen. However, this combination is impractical to use in any operational rocket. Lithium and fluorine are both extremely corrosive, lithium ignites on contact with air, fluorine ignites on contact with most fuels, and hydrogen, while not hypergolic, is an explosive hazard. The rocket exhaust is also ionized, which would interfere with radio communication with the rocket (https://en.wikipedia.org/wiki/Specific_impulse - cite_note-37). Hence, this is not in use, and it has only theoretical interest.

INCREMENTAL VELOCITY – THE IDEAL ROCKET EQUATION (REF. NASA)

The forces on a rocket change dramatically during a powered flight, as the propellants are constantly being exhausted from the nozzle. As a result, the mass of the rocket is constantly decreasing. Because of the changing mass, the standard form of Newton's second law of motion to determine the acceleration and velocity of the rocket, cannot be used. The following figure shows a derivation of the incremental velocity during powered flight, while accounting for the changing mass of the rocket. In this derivation, the effects of aerodynamic lift and drag are neglected.

M is the instantaneous mass of the rocket, u is the velocity of the rocket, v is the velocity of the exhaust from the rocket, A is the area of the exhaust nozzle, p is the exhaust pressure, and p0 is the atmospheric pressure. During a small amount of time dt, a small amount of mass dm is exhausted from the rocket. The "a" is the angle to the flight path.

From the changes in the mass and the velocity of the rocket, we can arrive at the change in momentum of the rocket as $M (u + du) - M u = M du$. We can also determine the change in momentum of the small mass dm that is exhausted at velocity v as $= dm (u - v) - dm u = - dm v$

So the total change in momentum of the system (rocket + exhaust) is

$$= \textbf{M du} - \textbf{dm v}$$

Now consider the forces acting on the system, neglecting the drag on rocket. The weight of the rocket is M g (gravitational constant) acting at an angle a to the flight path. The pressure force is given by $(p - p_o) A$ acting in the positive u direction. Then the total force on the system is $\textbf{(p} - \textbf{p}_o\textbf{) A} - \textbf{M g cos (a)}$

The change in momentum of the system is equal to the impulse on the system, which is equal to the force on the system times the change in time dt. So we can combine the previous two equations:

$$M \, du - dm \, v = [(p - p_0)A - M \, g \, cos(a)] \, dt$$

If we ignore the weight force, and perform a little algebra, this becomes

$$M \, du = [(p - p_0)A]dt + dm \, v$$

Now, the exhaust mass dm is equal to the mass flow rate mdot times the increment of time dt. So, we can write the last equation as

$$M \, du = [(p - p_0) A + mdot \, v]dt$$

On the web page describing the specific impulse, we introduce the equivalent exit velocity Veq, which is defined as

$$Veq = v + (p - p_0) * A/mdot$$

If we substitute the value of Veq into the momentum equation, we have

$$M \, du = Veq \, mdot \, dt$$

mdot dt is the amount of change of the instantaneous mass of the rocket. The sign of this term is negative because the rocket is losing mass as the propellants are exhausted.

$$mdot \, dt = -dM$$

Substituting into the momentum equation:

$$M \, du = -Veq \, d \, M$$

$$du = -Veq \, dM/M$$

We can now integrate this equation:

$$\textbf{delta u} = -\textbf{Veq ln(M)}$$

where delta u represents the change in velocity, and ln is the symbol for the natural logarithmic function. The limits of integration are from the initial mass of the rocket to the final mass of the rocket. The instantaneous mass of the rocket M, the mass is composed of two main parts, the empty mass me, and the propellant mass mp. The empty mass does not change with time, but the mass of propellants on board the rocket does change with time:

$$M(t) = me + mp(t)$$

Initially, the full mass of the rocket mf contains the empty mass and all of the propellant at lift off. At the end of the burn, the mass of the rocket contains only the empty mass:

$$M \, initial = mf = me + mp$$

$$M \, final = me$$

Substituting for these values we obtain:

$$\textbf{delta u} = \textbf{Veq ln(mf/me)}$$

This equation is called the ideal rocket equation. There are several additional forms of this equation which we list here: Using the definition of the propellant mass ratio MR

$$MR = mf/me$$

$$delta \, u = Veq * ln(MR)$$

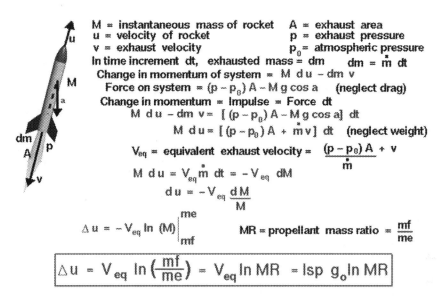

Veq is related to the specific impulse Isp:

$$Veq = Isp * g_0$$

where g_0 is the gravitational constant. So, the change in velocity can be written in terms of the specific impulse of the engine:

$$delta\ u = Isp * g_0 * ln(MR)$$

If we have a desired delta u for a manoeuver, we can invert this equation to determine the amount of propellant required:

$$MR = exp(delta\ u/(Isp * g_0))$$

where exp is the exponential function.

If you include the effects of gravity in the rocket equation, the equation becomes:

$$delta\ u = Veq\ ln(MR) - g_0 * tb$$

where tb is the time for the burn.

Valid Assumptions for an Ideal Rocket
(ref. Rocket Propulsion Elements by George H Sutton and Oscar Biblarz)

The concept of ideal rocket propulsion systems is useful because the relevant basic thermodynamic principles can be expressed as simple mathematical relationships. These equations theoretically describe a quasi-one-dimensional nozzle flow, which corresponds to an idealization and simplification of the full two- or three-dimensional equations and the real aero thermochemical behaviour. In designing new rockets, it has become accepted practice to use ideal rocket parameters which can then be modified by appropriate corrections. An ideal rocket unit is one for which the following assumptions are valid:

1. **The working substance (or chemical reaction products) is homogeneous.**
2. **All the species of the working fluid are gaseous. Any condensed phases (liquid or solid) add a negligible amount to the total mass.**
3. **The working substance obeys the perfect gas law.**
4. **There is no heat transfer across the rocket walls; therefore, the flow is adiabatic.**
5. **There is no appreciable friction, and all boundary layer effects are neglected.**
6. **There are no shock waves or discontinuities in the nozzle flow.**
7. **The propellant flow is steady and constant. The expansion of the working fluid is uniform and steady, without vibration. Transient effects (i.e., start up and shut down) are of very short duration and may be neglected.**
8. **All exhaust gases leaving the rocket have an axially directed velocity.**
9. **The gas velocity, pressure, temperature, and density are all uniform across any section normal to the nozzle axis.**
10. **Chemical equilibrium is established within the rocket chamber, and the gas composition does not change in the nozzle (frozen flow).**
11. **Stored propellants are at room temperature. Cryogenic propellants are at their boiling points.**

STAGING/MULTISTAGE ROCKETS

The purpose of the satellite launch vehicle is to inject a reasonably large payload in the specified orbit with required velocity. A single stage to carry out such mission calls for huge, massive rocket. Therefore, the designers choose multistage rocket, throwing away the burnt-out empty stage from the lower end, in sequence. Discarding the structural mass helps in building the incremental velocity at every stage to obtain the needed velocity at the last upper stage. This concept is called staging.

Staging can be either serial, one over the other or can be parallel as strap-ons in the booster system. ISRO rockets SLV-3 is an example for the former and GSLV-Mk III is an example for parallel staging, as in Figure 4.10.

SLV-3 has four stages with solid propellant rocket motors. The first stage weighs 10 tons with a propellant mass of 9 tons, giving a peak thrust of 54 t to lift a weight of 17 t. The second stage weighs only one third, generating thrust of 29 t to lift 7 t mass, increasing the velocity. The fourth stage injects the Rohini satellite with 7.88 km/s orbital velocity.

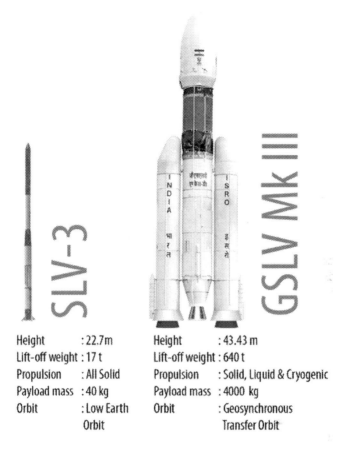

Height   : 22.7 m
Lift-off weight : 17 t
Propulsion   : All Solid
Payload mass : 40 kg
Orbit   : Low Earth
               Orbit

Height   : 43.43 m
Lift-off weight : 640 t
Propulsion   : Solid, Liquid & Cryogenic
Payload mass   : 4000 kg
Orbit   : Geosynchronous
                Transfer Orbit

FIGURE 4.10 Specifications of SLV-3 and GSLV Mk III.

GSLV Mk III is designed to carry 4 ton class of satellites into Geosynchronous Transfer Orbit (GTO) or about 10 tons to Low Earth Orbit (LEO).

The two strap-on motors of GSLV Mk III are located on either side of its core liquid booster. Designated as "S200," each carries 205 tons of composite solid propellant and their ignition results in vehicle lift-off with 640 t mass, with lift off thrust of 1000 t. S200s function for 140 seconds. During strap-ons functioning phase, the two clustered Vikas liquid Engines of L110 liquid core booster will ignite 114 seconds after lift-off to further augment the thrust of the vehicle. These two engines continue to function after the separation of the strap-ons at about 140 seconds after lift-off.

5 Rocket Systems Development

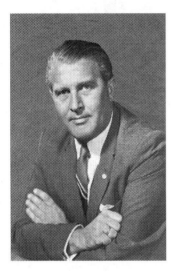

The fire-arrows of the Chinese and the first metal rockets ever built in India and later re-engineered by the British in those days must have been exciting, but it was a highly dangerous activity. Many just exploded on launching. Others flew on erratic courses and landed in wrong places, due to absence of control.

Today, rockets are big, efficient, and much more reliable. They fly on precise courses to orbit satellites and can go fast enough to escape the gravitational pull of Earth. Our understanding of the scientific principles behind rocketry has led us to develop newer advanced propulsion, guidance and control, software, and testing methodologies that can be used for highly efficient and more powerful rockets for different exciting missions, re-entry, and re-use. Let us discuss the anatomy of a typical rocket.

ROCKET SUBSYSTEMS AND COMPONENTS

A rocket is an integrated system and has various subsystems, components, and interconnectivities for different functions to achieve its objective of placing a spacecraft, crew, or cargo in a desired orbit with required injection parameters. There are four major components to a rocket to meet its defined goal – viz. propulsion system, structure or frame, guidance & control, and payload. The breakdown structure is shown in Figure 5.1.

DOI: 10.1201/9781003323396-5

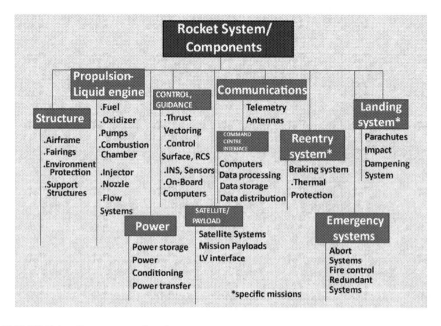

FIGURE 5.1   Components of rocket system.

The components of rocket and their functions are:

- **Propulsion:** (Liquid engine) – fuel & oxidiser, pump, flow systems, injector, combustion chamber, nozzle, and others needed for propelling the vehicle. (Solid motor) – solid propellant, insulation, igniter, and nozzle.
- **Structure:** Rocket motor casings/ tankages, inter stages, support structure for all the components, fairings with appropriate aerospace materials (metallic or composites) and fabrication processes.
- **Control & Guidance:** Different control systems with stages – attitude control system (ACS) and reaction control systems; thrust vector controls (TVC) and control surfaces, such as fins or wings; navigational sensors, such as inertial guidance units and star trackers, and on-board computer loaded with mission software, electronics.
- **Power:** Power storage (battery), conditions that power for use, and distribution, harnessing.
- **Telemetry electronics** with command computers, data processing, antenna.
- **Satellite/Payload:** Spacecraft with mission instruments and subsystems, crew, cargo, launch vehicle-satellite interface. Warhead in the case of a missile.
- **Thermal protection system for re-entry**, and recovery with parachutes, braking system, impact dampening system (for specific missions).
- **Communications**, control centre interface, data processing.
- **Abort systems**, redundant systems, umbilical interface with launcher.

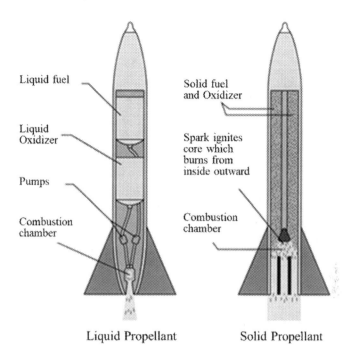

Liquid fuel

Liquid
Oxidizer

Pumps

Combustion
chamber

Solid fuel
and Oxidizer

Spark ignites
core which
burns from
inside outward

Combustion
chamber

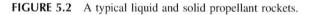

Liquid Propellant   Solid Propellant

FIGURE 5.2   A typical liquid and solid propellant rockets.

A typical rocket with all subsystems – liquid and solid propellant rockets – is shown in Figure 5.2:
Let us detail the significance of each system.

PROPULSION

The general classification of propulsion systems is shown in Figure 5.3.

SOLID PROPELLANT ROCKET

A rocket is different from a jet engine. A jet engine requires oxygen from the air to work. A rocket engine carries everything it needs. That is why a rocket engine works in space, where there is no air.

There are two main types of rocket engines. Some rockets use solid fuel. SLV-3, Agni missile, strap-on boosters of Ariane, and H-II use solid fuel- propellant. Solid fuel rockets are used as boosters to give huge take-off thrust, like the solid rocket boosters (SRBs) of the space shuttle.

LARGE SOLID PROPELLANT BOOSTER

The SRBs of the space shuttle provide 71.4% of its thrust during liftoff and ascent, and are the largest reusable solid-propellant motors. Each SRB is 45 m tall and

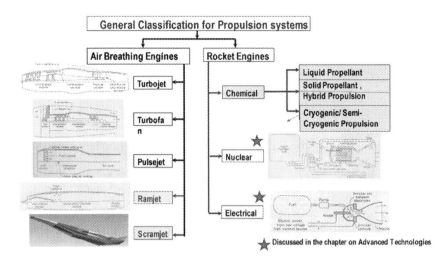

FIGURE 5.3   General classification for propulsion systems.

3.7 m wide, weighs 68,000 kg, and has a high-strength steel exterior 13 mm thick. The SRB's subcomponents are the solid-propellant motor, nose cone, and rocket nozzle. The solid-propellant motor comprised the majority of the SRB's structure. Its casing consists of 11 steel sections, which make up its four main segments, with tongue and screw joints. The nose cone houses the forward separation motors and the parachute systems that were used during recovery. The rocket nozzles could gimbal up to 8° to allow for in-flight adjustments.

Each of the rocket motors is filled with a total 500,000 kg of solid rocket propellant, providing 13,300 kN thrust. The propellant is based on PBAN (Poly Butadiene Acrylo Nitrile) binder (12%), with atomised aluminium powder as fuel (16%) and APCP (Ammonium Per Chlorate Powder) as oxidiser (69.8%). Iron oxide of 0.2% is added to assist the combustion process. In addition to providing thrust during the first stage of launch, the SRBs provide structural support for the orbiter vehicle and external tank, as they are the only system that is connected to the mobile launcher platform (MLP). During the space shuttle program (which is closed now), the SRBs were jettisoned two minutes after launch of space shuttle at an altitude of approximately 46 km, after expending their fuel. Following separation, the boosters deployed drogue and main parachutes, landed in the ocean, and were recovered by the crews aboard the ships. Once they were returned to Cape Canaveral, cleaned, and disassembled for refurbishment.

A solid rocket motor is shown in Figure 5.4, with its structure and components.

PBAN (Polybutadiene Acrylo Nitrile) and HTPB (Hydroxyl Terminated Polybutadiene) are the commonly used binders for the composite propellant along with Al powder, additives, and the oxidiser ammonium perchlorate (APCP). The propellant grain of typical solid rocket motor makes up about 85%–90% of the rocket motor's total mass.

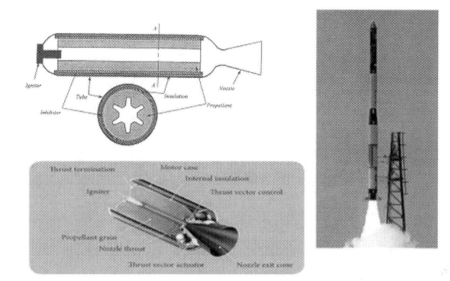

FIGURE 5.4   Solid propellant rocket SLV-3.

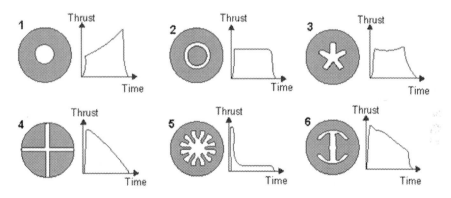

FIGURE 5.5   Some of the grain geometries and their corresponding thrust curves.

Some motors have a star-shaped channel along the central axis of the rocket, whereas the wall of the inside open channel is the burning surface, shown in Figure 5.5. Some grains have no hole in them, and they are called end- burning grain. Here the grain burns from one end. Others have more exotic burning surfaces, depending on grain design, based on mission requirement (thrust-time profile-progressive (1), neutral (2) or regressive (6).

BURN RATE

Depending on the composition of the propellant, grain configuration and resultant chamber pressure, the propellant burns at a rate, given in mm per second. The burn

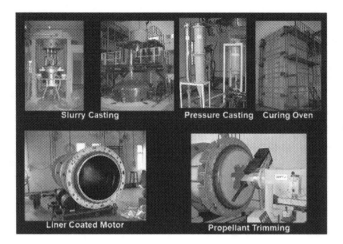

FIGURE 5.6   Advanced propellant processing of rocket motor.

rate is can be calculated by the formula (Saint- Robert's law) $r = a\,P^n$, where r is the burn rate, a is the temperature coefficient, P is the pressure of rocket motor and n is the combustion index.

Exterior to the grain is thermal insulation barrier. This barrier protects the outer casing of the motor from the extreme temperatures and pressures of the rocket motor. Insulation rubber such as Rocasin (rocket case insulation developed by ISRO) or EPDM (ethylene propylene diene monomer – a type of synthetic rubber) is normally used for solid rocket motors. The casing is typically the only part of a solid motor that can be reused, if designed for multi-use. The process of preparation of the motor with casing readiness, insulation lining, propellant casting with mandrel to give grain shape, curing in oven, and end trimming are shown in Figure 5.6.

The part of the solid motor within the casing that houses the grain, and the burning surface is the combustion chamber. Solid rocket motors are designed in various shapes and sizes to optimise the combustion chamber for efficient burning of the propellant and for generating the desired thrust-time profile, as seen earlier. At the bottom of the combustion chamber the inlet to the convergent–divergent nozzle is connected where the flow is accelerated out of the motor to generate the desired thrust. This is proved on ground by static testing of the motor, to establish the thrust-time profile in flight.

SLV-3 was designed with four solid propellant rocket motors. During development we had to go through 56 static tests in all, before the motors were declared fit for flight. The four solid propellant rocket motors of SLV-3 can be seen in the exploded view shown in Figure 5.7.

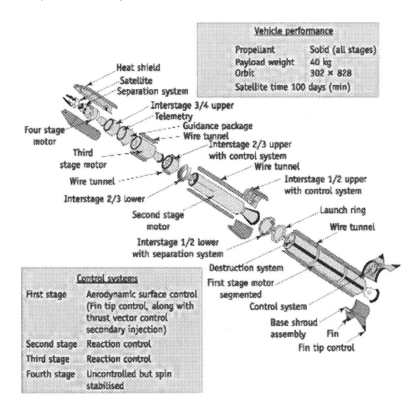

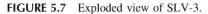

FIGURE 5.7   Exploded view of SLV-3.

LIQUID PROPELLANT ROCKETS

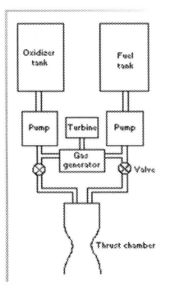

Vikas Engine (Courtesy ISRO):   Ground Testing

FIGURE 5.8 Liquid propellant rocket.

Liquid-propellant rockets have four principal components in their propulsion system: propellant (fuel and oxidizer) tanks, the rocket engine's combustion chamber and nozzle assembly, gas generator, and turbo pumps, as shown. The fuel and oxidizer are stored in separate tanks, as they are hypergolic. Some of the propellant is burned in the gas generator, and the resulting hot gas is used to power the turbo pumps. The liquid propellants are then pumped into the combustion chamber, by the two separate turbo pumps in required quantity, through an injector in the form of fine vapour spray, based on the most efficient stoichiometric ratio. Being hypergolic in nature, combustion takes place instantaneously, creating the hot exhaust gases, which then expand through the nozzle, generating thrust. Vikas engine used in PSLV, as second stage is shown as example of liquid propellant rocket in Figure 5.8.

Vikas engine generates 800 kN thrust, operating at 58.5 bar pressure. The design was based on the licensed version of the Viking engine of SEP, France. This uses turbo pump fed system, as explained above.

Pressure-Fed System

The propellant feed system of a liquid rocket engine determines how the propellants are delivered from the tanks to the thrust chamber. We discussed in earlier para the turbo pump feed system for Vikas engine. The other type is called pressure-fed system, which is simple and relies on tank pressures to feed the fuel and oxidizer into the thrust chamber (Figure 5.9). This type of system is typically used for satellite propulsion and auxiliary propulsion applications requiring low system pressures and small quantities of propellants to operate thrusters of few newtons.

The pressure-fed system generally includes: 1) pressurised tank(s) to store propellants, 2) pressurant gas or other expulsion device to provide energy for the feed system, 3) valves to control the pressure and flow, and 4) ducting or piping to transfer fluids to thrust chamber to generate thrust. It can be as simple as a cold gas

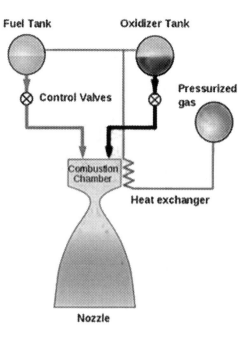

FIGURE 5.9   Pressure fed system of liquid propellant engine.

Courtesy: astronomicalretu.

thruster, which has a pressurised tank connected to a propellant tank either 1) monopropellant or 2) bipropellant systems. The monopropellant system uses a single propellant (e.g., hydrazine) flowing through a catalyst-bed prior to expanding in a converging-diverging nozzle. The pressure fed bi-propellant systems, which employ an oxidizer tank and a fuel tank, both of which require a pressurant system to expel the propellant into the feed system. In the case of SLV-3 the second stage control system used RFNA and Hydrazine as oxidizer and fuel respectively. The third stage control system was monopropellant – Hydrazine. Titanium alloy high pressure air bottles provide required pressure to push the propellants to flow.

The pressure-fed systems are primarily used for orbit manoeuvre, orbit insertion, attitude control, reaction control, and small upper stage propulsion.

TYPES OF LIQUID PROPULSION

There are three types of liquid propellants used in rockets-earth storable, semi-cryogenic and cryogenic.

1. Storable propellant combinations are hypergolic in nature and used in large rocket engines for first and/or second stages. Also, in almost all bipro-pellant low-thrust, auxiliary or reaction control rocket engines. Storable liquid propellants nitrogen tetroxide (N_2O_4) and unsymmetrical dimethyl hydrazine (UDMH) are generally used.

In the case of Prithvi missile, RFNA (Red Fumic Nitric Acid which is a combination of HNO_3 + NO_2) and G-Fuel (equimolar mixture of Xylidine and Triethylamine) are used as propellants. They are corrosive in nature and any leakage must be treated immediately.

These propellants allow reasonable time of storage, and almost instant readiness to start without delays. In many satellites and upper stages bipropellant thrusters are used with nitrogen tetroxide and mono methyl hydrazine (MMH) to give service for more than ten years. Alternatively, satellites have also used monopropellant hydrazine for auxiliary engines/thrusters for station keeping.

As the liquid propellant rocket is giving required performance, the possible problems such as pogo, slash and cavitation in the blades of turbo pump could arise. Necessary corrective measures are taken during design. Cooling of the engine is another area to be looked at for desired functioning of the engine.

2. The liquid oxygen with hydrocarbon propellant (kerosene, propane, or methane) combination (semi-cryogenic), used for booster stages of space launch vehicles; due to higher specific impulse and higher average density, when compared to the storable propellants. Historically, semi-cryogenic rockets were the first developed and was originally used for ballistic missiles in Russia.

3. Cryogenic oxygen-hydrogen (Lox-LH2) is used in upper stages of space launch vehicles, due to higher specific impulse of the combination, especially for high vehicle velocity missions. In the design of heavy lift launch vehicles like space shuttles, Ariane-5, and H-II, Lox-LH2 combination was/is being used as large core first stage with solid strap-on boosters to give necessary lift.

CRYOGENIC ROCKET ENGINES (LOX + LH2)

In a **cryogenic rocket engine** both its fuel and oxidizer are gases liquefied and stored at very low temperatures. Various cryogenic fuel-oxidizer combinations have been tried, but the combination of liquid hydrogen (LH2) fuel and the liquid oxygen (LOX) oxidizer is one of the most widely used. Both are easily and cheaply available, and when burnt have one of the highest enthalpy releases in combustion, producing a specific impulse of 450 seconds at an effective exhaust velocity of 4.4 km per second. These highly efficient engines were first flown on the US Atlas-Centaur (RL-10).

Rocket engines burning cryogenic propellants remain in use today on high performance upper stages and boosters. United States, Russia, Japan, France, China and India have cryogenic engines and stages. ESA's Ariane 5, JAXA's H-II, and the United States Space Launch System use cryogenic core with solid strap-on boosters to augment take-off thrust.

While it is possible to store propellants as pressurised gases, this would require large, heavy tanks that would make achieving orbital spaceflight difficult if not impossible. On the other hand, if the propellants are cooled sufficiently, they exist in the liquid phase at higher density and lower pressure, simplifying tankage. These cryogenic temperatures vary depending on the propellant, with liquid oxygen existing below −183°C and liquid hydrogen below −253°C.

The major components of a cryogenic rocket engine are the combustion chamber, pyrotechnic initiator, fuel injector, fuel and oxidizer turbo pumps, cryogenic valves, regulators, the fuel tanks, and rocket engine nozzle. In terms of feeding propellants to the combustion chamber, cryogenic rocket engines are almost exclusively pump-fed. Pump-fed engines work in a gas-generator cycle, a staged-combustion cycle, or an expander cycle. Figure 5.10 explains engine, typical testing, and application of cryogenic engine in GSLV.

A typical rocket using all three types of propulsion is GSLV Mk II.

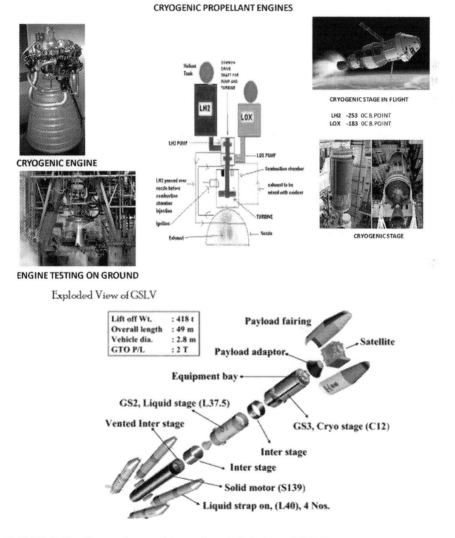

FIGURE 5.10   Cryogenic propulsion and exploded view of GSLV.

Semi-cryogenic Rocket Engines

Semi-cryogenic engine uses liquid oxygen (LOX) as oxidiser with RP-1 kerosene or methane or propane as fuel. In early rockets Soviets used extensively aviation kerosene (RP-1) as fuel with liquid oxygen as oxidizer, for their large boosters. Kerosene is lighter than liquid hydrogen and can be stored in normal temperature. Kerosene combines with liquid oxygen provide a higher thrust to the rocket, compared to solid and storable liquid propellants. Moreover kerosene occupies lesser space, making it possible to carry more propellant in semi-cryogenic engines fuel tankages. A semi-cryogenic engine is environment friendly and cost-effective compared to cryogenic engine, and ideal for boosters of launch vehicles. Vostok, in early 1960s used LOX-RP1 engines, 16 of them in four boosters, each generating a thrust of 970 KN, and core first stage with four engines. The upper stage was also semi-cryogenic engine of smaller version, making all through semi-cryogenic propulsion.

Saturn V for Apollo mission used five semi-cryogenic F-1 engines using RP-1 kerosene and Lox, each generating a sea level thrust of 6670 KN. The F-1 remains the most powerful single combustion chamber liquid-propellant rocket engine ever developed (Figure 5.11).

SpaceX uses Merlin family of rocket engines on its Falcon 1, Falcon 9 and Falcon Heavy launch vehicles. Merlin engines use RP-1 and liquid oxygen as rocket propellants in a gas-generator power cycle. Propellants are fed by a single-shaft, dual-impeller turbo-pump. The turbo-pump also provides high-pressure fluid for the hydraulic actuators, which then recycles into the low-pressure inlet. This eliminates the need for a separate hydraulic drive system. The engine generates 845 KN thrust (Figure 5.12).

Raptor is a family of full-flow staged combustion cycle rocket engines developed and manufactured by SpaceX, for use on the in-development Star-ship fully reusable launch vehicle. The engine is powered by cryogenic liquid methane and liquid oxygen (LOX), rather than the RP-1 kerosene and LOX used in SpaceX's prior Merlin and Kestrel rocket engines. The Raptor engine has more than twice the thrust of SpaceX's Merlin engine that powers their current Falcon 9 and Falcon Heavy launch vehicles.

FIGURE 5.11   Semi-cryogenic F-1 engine.

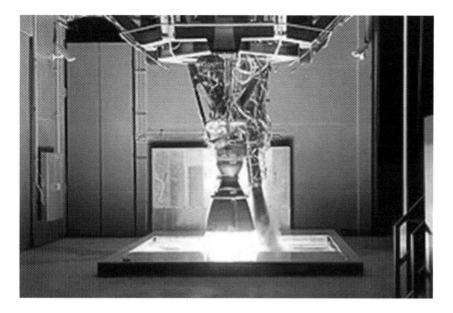

FIGURE 5.12   Test firing of the merlin 1D at SpaceX's McGregor.

ISRO SEMI-CRYOGENIC ENGINE (SCE-200)

India and Ukraine have been collaborating closely since the signing of the Framework Agreement for cooperation in 2005. India's semi-cryogenic engine, the SCE-200, will have 200 tonnes of thrust, and will also be using kerosene and LOX. The SCE-200 will use an oxidizer-rich staged combustion cycle. The engine will be tested in collaboration with the Ukrainian firm Yuzhmash, and once it is ready it will be used to equip future launchers of ISRO like the Unified Launch Vehicle (ULV), which is currently in development. This 200 t core stage will replace the present 110 t storable liquid propellants UDMH+N_2O_4 in the GSLC Mk-III. This will enable lifting 6 t communication satellite to GTO, from the present capacity of 4 t by GSLV Mk-III.

A combination of solid and liquid fuels is also used in some applications. In its simplest form, a hybrid rocket consists of a pressure vessel (tank) containing the liquid oxidisers, the combustion chamber containing the solid propellant, and a mechanical device separating the two. When thrust is desired, a suitable ignition source is introduced in the combustion chamber and the valve is opened. The liquid oxidizer (or gas) flows into the combustion chamber, where it is vaporised and then reacted with the solid propellant. Combustion occurs in a boundary-layer diffusion flame adjacent to the surface of the solid propellant.

Generally, the liquid propellant is the oxidizer, and the solid propellant is the fuel, because solid oxidisers are extremely dangerous and lower performing than liquid oxidisers.

So far, we went through rockets that use either liquid or solid propellants, which are burnt within a high-pressure chamber to produce hot gaseous products that are expanded through an exhaust nozzle to produce thrust. They can function in any altitude, including above atmosphere. But air-breathing engines can only work within the limits of atmosphere, as air is the oxidizer. Air-breathing propulsion systems can be categorised as turbojets, ramjets, ducted rockets, scramjets, and the dual-combustion ramjet (DCR). A ducted rocket where the fuel rich effluent of a rocket motor is mixed in a downstream combustor with air captured from the atmosphere to improve the efficiency of the engine cycle. Conventional hydrocarbon fuels can be used. In addition, they can be throttled to allow trajectory flexibility. Some examples of systems being operated in different sonic regimes are shown in Figure 5.13.

Turbojet: Conventional turbojet engines use mechanical compression in the inlet (cold), and airstream compression by a turbine located downstream (hot) (Figure 5.14a).

Ramjet: Ramjets can operate efficiently at supersonic speeds with sufficient compression in the inlet at subsonic speed. With exhaust speed more than Mach 1, the engine operates supersonically. Ramjet can operate up to Mach 5. Above Mach 5 it is Scramjet. Both the ramjet and scramjet must be coupled with some additional form of propulsion to accelerate the vehicle to the required speed (Figure 5.14b).

The action of rocket propulsion systems and ramjets can be combined, utilising a common combustion chamber. Such a low-volume configuration, known as an integral rocket–ramjet, has been attractive for use in supersonic missiles.

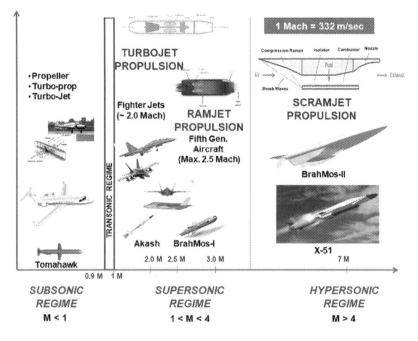

FIGURE 5.13 Sonic regimes.

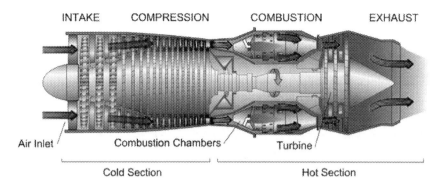

FIGURE 5.14a Turbojet engine.

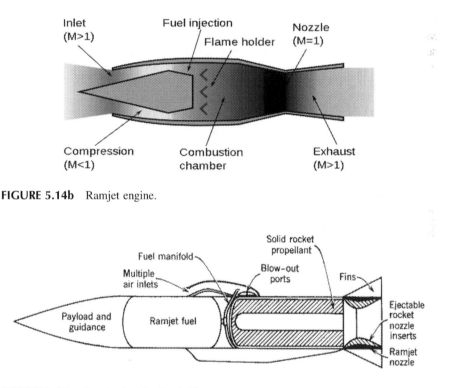

FIGURE 5.14b Ramjet engine.

FIGURE 5.14c Ramjet in Akash missile.

Figure 5.14c shows a missile with integral rocket–ramjet propulsion. After the solid propellant has been consumed in boosting the vehicle to flight speed of about Mach 2, the rocket combustion chamber becomes the ramjet combustion chamber with air burning the ramjet fuel (fuel rich boron based solid or RP-7 aviation kerosene). This combines the principles of rocket and ramjet engines giving higher performance (specific impulse) than chemical rocket engines. But it can only operate within the earth's atmosphere.

Akash is a surface-to-air missile with an intercept range of 30 km. It has a launch weight of 720 kg, a diameter of 35 cm and a length of 5.78 m. Akash flies at supersonic speed, reaching around Mach 2.5. It can reach an altitude of 18 km and can be fired from both tracked and wheeled platforms. An on-board guidance system coupled with an actuator system makes the missile manoeuvrable up to 15 g loads and a tail chase capability for end game engagement. A digital proximity fuse is coupled with a 55 kg pre-fragmented warhead, while the safety arming and detonation mechanism enables a controlled detonation sequence. It is propelled by an Integrated Ramjet Rocket Engine. The use of a ramjet propulsion system enables sustained speeds without deceleration throughout its flight.

The missile has four long tube ramjet inlet ducts mounted mid-body between wings. For pitch/yaw control four clipped triangular moving wings are mounted on the mid-body. For roll control four inline clipped delta fins with ailerons are mounted before the tail. Composite technology for Akash includes radome assemblies, booster liners, ablative liners, sustainer liners, compression moulded wings and fins.

LIQUID RAMJET PROPULSION

BRAHMOS supersonic cruise missile uses liquid ramjet with solid propellant booster. BrahMos is powered by a two-stage propulsion system. Initial acceleration is provided by a solid-propellant booster and supersonic cruise speed is provided by a liquid-fuelled ramjet system. The air-breathing ramjet propulsion is more fuel-efficient in comparison with conventional rocket propulsion. It provides the BrahMos with a longer range over similar missiles powered by rocket propulsion.

The ramjet is designed around its inlet (Figure 5.15). An object moving at high speed through air generates a high-pressure region upstream. A ramjet uses this high pressure in front of the engine to force air through the tube, where it is heated by combusting some of it with fuel. It is then passed through a nozzle to accelerate it to supersonic speeds.

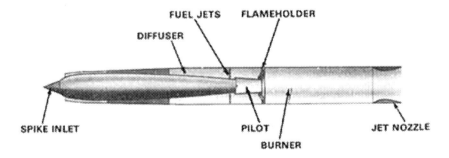

FIGURE 5.15 Liquid ramjet in BRAHMOS missile.

Ramjet utilises high-speed characteristics of air to literally "ram" air through an inlet diffuser into the combustor. At transonic and supersonic flight speeds, the air upstream of the inlet is not able to move out of the way quickly enough and is compressed within the diffuser before being diffused into the combustor. Combustion in a ramjet takes place at subsonic velocities, but the combustion products are then accelerated through a convergent-divergent nozzle to supersonic speeds. A ramjet is sometimes referred to as a "flying stovepipe", a very simple device comprising an air intake, a combustor, and a nozzle. Normally, the only moving parts are those within the turbo pump, which pumps the fuel to the combustor in a liquid-fuel ramjet.

The energy level for each of the propulsion systems is measured in terms of Specific Impulse (Second). Solid and liquid propellants provide 300 seconds, Cryogenics give 450 seconds, Solid ramjet gives 600 seconds, and the liquid ramjet gives 1200 seconds of specific impulse along with the possible speed of the system in Mach number.

Liquid ramjet engine has a definite advantage of four times the energy level of solid propellants in addition to low weight, low volume, excellent fuel economy and a wide range of altitude of operation in supersonic cruise mode and also longtime storability. Utilising the availability of the liquid ramjet engine with the NPO Mashinostroyenia, Russia, DRDO decided to work with them to evolve a feasible jointly designed cruise missile. That is BrahMos, named after the two rivers Brahmaputra and Moskva of India and Russia.

The cruise missile is a self-propelled guided airborne vehicle which has aerodynamic lift due to its wings and continuous operation propulsion throughout its flight. The advantages of cruise missiles are (a) control of thrust during its flight, (b) operability at a wide range of altitudes, (c) compactness in size giving minimum radar cross section signature, (d) excellent fuel economy as the oxidizer is from atmosphere, and (e) high precision. In addition, if it has supersonic speed, it will give tremendous effect in destruction of the target, apart from minimum reaction time given to the enemy to defend himself.

BrahMos is a canisterised missile and can be launched from ship, land, silo, submarine, and aircraft. The accuracy and speed of the missile and versatility of launch platforms make it most lethal. It has a first stage of solid propellant and a second stage of liquid ramjet engine. It is guided by inertial navigation during the course of its flight and during the terminal phase it locks on the target with the help of an on-board seeker. It is a tactical missile and has highly manoeuvrable trajectories.

SUPERSONIC-COMBUSTION RAMJET (SCRAMJET)

A scramjet engine is an improvement over the ramjet engine as it efficiently operates at hypersonic speeds, more than Mach 5 and allows supersonic combustion. Thus, it is known as Supersonic Combustion Ramjet, or Scramjet

Ramjets always slow the incoming air to a subsonic velocity within the combustor. In scramjet, the air goes through the entire engine at supersonic speeds. This increases the stagnation pressure recovered from the free stream and improves net thrust. Thermal choking of the exhaust is avoided by having a relatively high supersonic air velocity at combustor entry. Fuel injection is often into a sheltered region below a step in the combustor wall, as shown in Figure 5.16.

The scramjet is composed of three basic components: 1) a converging inlet, where incoming air is compressed; 2) a combustor, where gaseous fuel is burned with atmospheric oxygen to produce heat; and 3) a diverging nozzle, where the heated air is accelerated at high speed to produce thrust. Unlike a typical jet engine, such as a turbojet or turbofan engine, a scramjet does not use rotating, fan-like components to compress the air.

Due to the nature of their design, scramjet operation is in hypersonic velocities. As they lack mechanical compressors, scramjets require the high kinetic energy of a hypersonic flow to compress the incoming air to operational conditions. Thus, a

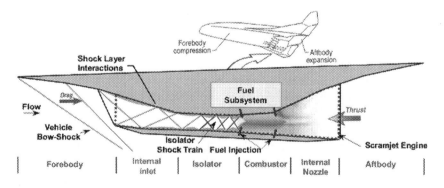

FIGURE 5.16 Supersonic combustion ramjet (SCRAMJET).

scramjet-powered vehicle must be accelerated to the required velocity (usually about Mach 4) by some other means of propulsion, such as turbojet, railgun, or boost rocket or ramjet.

SCRAMJET RESEARCH

Conceptually scramjet is simple. But actual implementation is limited by extreme technical challenges. While scramjet technology has been under-development since the 1950s in some countries, only very recently have scramjets successfully achieved powered flight, that too for limited duration. Hypersonic flight within the atmosphere generates immense drag, and extreme temperatures. Maintaining combustion in the supersonic flow for longer time presents many challenges, as the fuel must be injected, mixed, ignited, and burnt within milliseconds. The current scramjet technology requires the use of high-energy fuels and active cooling schemes to maintain sustained operation, often using hydrogen and regenerative cooling techniques, and use of high temperature materials.

The Centre of Excellence established at IISc undertook focused collaborative research in high speed aerodynamics and associated interdisciplinary areas like Hypersonic flow control, new concepts in aerodynamic configuration, scramjet, numerical simulation and ground testing, supersonic gaseous mixing and related complex gas dynamics, special materials / coating for hypersonic flight, next generation control and guidance strategies for hypersonic speed regimes, special purpose MEMS and Nano sensor for hypersonic vehicles, chemical kinetics associated with fuels in scramjet engines, etc.

The following were the challenges for the Hypersonic Centre:

- Evolution of CFD Model for hypersonic flight system with the integrated flow studies from subsonic, sonic, supersonic and hypersonic systems
- Hypersonic flow control – proof of concept experiments in lab to actual applications
- Development of high lift generating surfaces
- Optimum configuration for SCRAMJET powered hypersonic speeds
- Complex gas dynamics at very hypersonic speeds
- Special materials/coatings for hypersonic speed regimes
- Next generation control and guidance for hypersonic vehicles
- Development of special purpose MEMS and Nanosensors
- Understanding of chemical kinetics associated with fuels in SCRAMJET engines.

Experimental flight tests have been carried out by ISRO and DRDO, for short duration. A lot more effort is needed to perfect the technology and configure a launch vehicle or missile.

This high speed makes the control of the flow within the combustion chamber more difficult. Since the flow is supersonic, no downstream influence propagates within the freestream of the combustion chamber. Throttling of the entrance to the thrust nozzle is not a usable control technique. In effect, a block of gas entering the combustion chamber must mix with fuel and have sufficient time for initiation and reaction, all the while traveling supersonically through the combustion chamber before the burnt gas is expanded through the thrust nozzle. This places stringent requirements on the pressure and temperature of the flow and requires that the fuel injection and mixing be extremely efficient. The emerging need is innovation, research, and development of technologies for reusable hypersonic flying vehicles. The demand from hypersonic fighter aircraft and hypersonic cruise missiles will be in the areas of hypersonic aerodynamics, propulsions, materials, and structures, new CFD codes and so on. This research centre would make India to realise many out-of-box research results of world class leading to re-usable cruise missile and multiuse single stage to orbit hyperplane.

MACH NUMBER VS. SPECIFIC IMPULSE

Mach 8 and specific Impulse of 1200 seconds can be achieved for kerosene-based ramjet, whereas Mach 20 and specific impulse of 3000 seconds can be achieved for hydrogen-based scramjet, as shown in Figure 5.17.

RE-USABILITY OF BOOSTER

The space shuttle was the first reusable spacecraft to carry people into orbit; launch, recover, and repair satellites; conduct cutting-edge research; and help build the International Space Station.

The space shuttle consists of the shuttle orbiter, two solid rocket boosters, and an expendable external tank. It takes off leaving much smoke and fire. The space

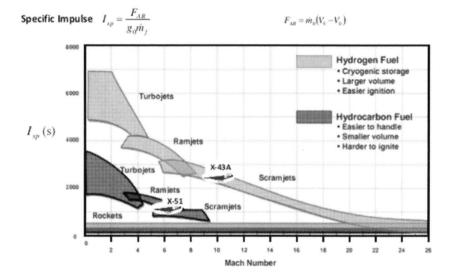

FIGURE 5.17   Hydrogen-based scramjet propulsion performance.

Credit: AIAA (SlideShare).

shuttle had a number of reusable parts. Solid fuel boosters on either side were recovered and refuelled after each flight, and the entire orbiter returned to Earth for use in subsequent flights. The large liquid fuel tank was expended. The space shuttle was a complex assemblage of technologies, employing both solid and liquid fuel and pioneering ceramic tiles as reentry heat shields. As a result, it permitted multiple launches as opposed to single-use rockets.

The long-term objectives of SpaceX include returning a launch vehicle first stage to the launch site in minutes and to return a second stage to the launch pad following orbital realignment with the launch site and atmospheric re-entry in up to 24 hours. SpaceX's long-term goal is that both stages of their orbital launch vehicle will be designed to allow reuse a few hours after return.

SpaceX first achieved a successful landing and recovery of a first stage. Second flights of refurbished first stages then became routine, with individual boosters having powered up to ten missions as of March 2021.

The reusable launch system technology was developed and initially used for the first stage of Falcon 9. After stage separation, the booster flips around, an optional boost-back burn is done to reverse its course, a re-entry burn, controlling direction to arrive at the landing site and a landing burn to affect the final low-altitude deceleration and touchdown.

STRUCTURE

In a rocket or launch vehicle, the structure holds integrity, keeping all functional systems and payload within. At the same time its mass must be minimum, for higher performance of the rocket. As a basic principle of rocket flight, it can be said that for a rocket to leave the ground, the engine must produce a thrust (T) that is greater than the weight of the vehicle (W). T/W must be more than 1. For an ideal (most recent) rocket SLS of NASA, the total mass of the vehicle is distributed as shown below. Of the total mass, 91% is propellants; 3% is structure like tanks, inter-stages, adapters, inside support frames both metallic and composites, heatshield, engines, fins, thrusters, electronics etc.; and 6% is the payload. Payloads may be satellites, astronauts, or spacecraft that will travel to other planets or moon.

Different structural parts of SLS rocket are shown in Figure 5.18.

MASS FRACTION

In determining the effectiveness of a rocket design, rocketeers speak in terms of mass fraction (MF). The mass of the propellants of the rocket divided by the total mass of the rocket gives mass fraction:

$$MF = (Mass\ of\ Propellants)/(Total\ Mass)$$

The mass fraction of the ideal rocket given above is 0.91. An MF number of 0.91 is a good balance between payload-carrying capability and range.

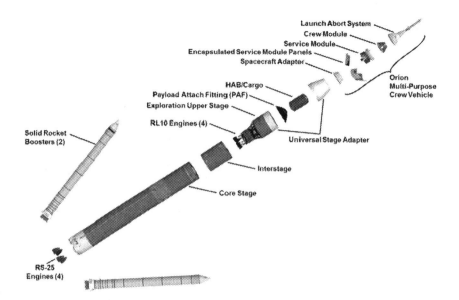

FIGURE 5.18 Exploded view of initial SLS configuration.

Courtesy: NASA.

Large rockets able to carry a spacecraft into space have serious weight problems. To reach space and find proper orbital velocities, a great deal of propellant is needed; therefore, the tanks, engines, and associated hardware become larger. Up to a point, bigger rockets fly farther than smaller rockets, but when they become too large their structures weigh them down too much, and the mass fraction is reduced to an impossible number.

As discussed earlier, when the large rocket motor was exhausted, the rocket casing was dropped behind and the remaining rocket fired. Much higher altitudes were achieved by this method, dropping the dead weight. The solid rocket boosters, strap-ons, and external tanks of larger rockets are dropped when they are exhausted of propellants.

As we saw in the chapter Principle of Rocket, its acceleration depends on three major factors, consistent with the equation for acceleration of a rocket.

First, the greater the exhaust velocity of the gases relative to the rocket, v (exit), the greater the acceleration is. The practical limit for v (exit) is about $2.5 \times 10^{*}3$ m/s for conventional (non-nuclear) hot-gas propulsion systems.

The second factor is the rate at which mass is ejected from the rocket (m dot). The faster the rocket burns its fuel, the greater its thrust, and the greater its acceleration.

The third factor is the mass m of the rocket. The smaller the mass is (all other factors being the same), the greater the acceleration. The rocket mass m decreases dramatically during flight because most of the rocket is fuel to begin with, so that acceleration increases continuously, reaching a maximum just before the fuel is exhausted.

MATERIALS

Every part of the rocket structure has special aerospace quality material, suitable to the function, keeping the mass minimum. The aim is to achieve higher structural strength and stiffness with lesser weight and higher propulsion efficiency, considering functional and environmental requirements. So, the need for exotic materials for aerospace applications as brought out in Figure 5.19. A few of the materials are discussed below. Rocket motor cases use high strength steel or titanium or composites.

A high-performance booster for PSLV (Figure 5.20(c)): For increased propellant loading and operating at high pressure, there was a need for very high strength steel, instead of the 15CDV6 steel used in the SLV-3. Maraging steel M250 was decided for PSLV booster knowing well its high fracture toughness.

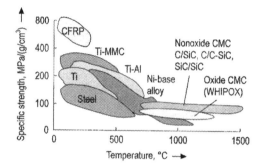

FIGURE 5.19 Exotic materials – Temperature vs. specific strength.

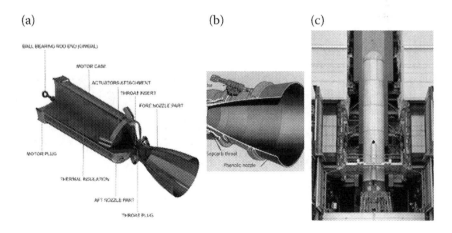

FIGURE 5.20 a) Rocket motor casing for solid propellant booster; b) Typical solid propellant nozzle structure; c) PSLV first Stage-S139 integrated.

Courtesy: ISRO.

This makes PSLV more reliable with lesser stages, instead of large strap-ons. MAR+AGING implies martensitic transformation in Fe-Ni matrix and its subsequent ageing in the presence of other alloying elements, in different grades 200, 250, and 300 depending on yield strength. The casings are fabricated by rolling and welding in five segments and sent for preparation with solid propellant as per the process already explained. Ablative liners and nozzle throat inserts are of composite materials such as carbon phenolic and silica epoxy layers to withstand high temperature.

UPPER STAGES

Weight of the structure being premium for upper stages composite materials such as Kevlar fibre is used, as in the case of PSLV third stage. Carbon composites casing is also used for similar application. The composite casings and structures used for upper stages provide remarkable reduction in structural weight to give high mass ratio for rocket motors. Carbon composites are used in satellite structure to reduce weight. Carbon-carbon, carbon-silicon carbide ceramic matrix composites are used in re-entry structures and nozzle throats to withstand temperatures of 2000 K–3000 K (Figure 5.21).

Titanium alloy Ti-6Al-4V, by virtue of its unique properties like high specific strength, good corrosive resistance to liquid propellants, good fracture toughness etc., is also used as rocket motor casing, especially in upper stages, as in PSLV fourth stage. This material is also used in high-pressure gas bottles.

(a)   (b)

FIGURE 5.21   a) PS-3 kevlar motor case, b) assembly of PS-3 with PS-4.

Courtesy: ISRO.

CASING FOR LIQUID PROPELLANT STAGES

High specific strength and specific modulus aluminum alloy – AA2014, 2219, 7075, and 6061 – material is good for ambient and cryogenic temperatures, high corrosion resistance property and more suitable for tankages to store fuel and oxidizer, and cryogenics. These materials can also be used for inter-stages, heatshield, and pressurising systems.

Magnesium alloys castings AZ92, ZE41, rolled AZ31, and ZK30 find extensive applications as secondary structures of launch vehicles and satellites, due to their high strength and less weight.

Copper impregnated Tungsten and Molybdenum based auto transpiration cooled composites are used for nozzle throat inserts of bi-propellant reaction control thrusters to withstand 2500°C temperature and highly oxidising conditions.

ALUMINAM-LITHIUM CASING FOR FALCON 9 (2010)

The Falcon launch system has recoverability and reusability of its first stage boosters. Reusability is a key element to SpaceX's aim of increasing the reliability and reducing the cost of spaceflight.

The first stage of the Falcon 9 launch system incorporates nine Merlin engines and aluminium-lithium alloy tanks containing liquid oxygen and rocket-grade kerosene (RP-1) propellant. Low-density Air ware 2195-T8 plate used in the tank barrels and domes of the booster. The combination of aluminium-lithium sheet and plate coupled with friction stir welding (FSW) enables the use of thin sections in this stability driven components within the rocket. The weight reduction achieved provides the possibility of an increased payload (about 10 tons to low earth orbit and 5 tons to geostationary orbit).

CONTROL SYSTEM

The rocket has a mission to travel in the prescribed trajectory and increase the velocity steadily to enable injecting the satellite at the correct altitude, inclination, velocity, and other orbit parameters. In order to impart force correction to steer in the preplanned trajectory (figure below) and to come back from the any deviated path the control system is necessary. The requirements for the control force are generated by the on-board computer based on sensors input, and the controls are ensured. There are several types of control systems. Small rockets usually require only a stabilising control system. Large rockets, such as the ones that launch satellites into orbit, require multiple systems in every stage rocket motors not only stabilise the rocket, but also enable it to change course while in flight. Large levels of shock induced by stage separations and deviations in propulsion performance need corrections by control systems.

Typical trajectory of launch vehicle injecting satellite & booster recovery – Falcon is shown in Figure 5.22.

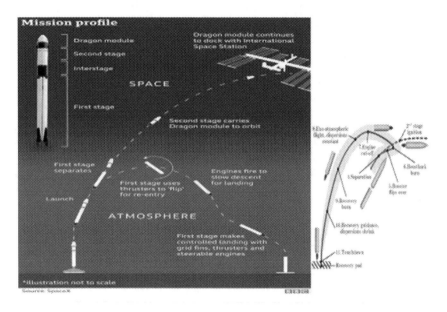

FIGURE 5.22   Typical trajectory profile, injection of satellite and booster recovery.

Types of Control

a. Engine Thrust Control

Controlling the thrust of an engine is very important to launching payloads (cargoes) into orbit. Too much thrust or thrust at the wrong time can cause a satellite to be placed in the wrong orbit or set too far out into space to be useful. Too little thrust can cause the satellite to fall back to Earth.

Liquid-propellant engines control the thrust by varying the amount of propellant that enters the combustion chamber. A computer in the rocket's guidance system determines the amount of thrust that is needed and controls the propellant flow rate. Complicated missions, such as going to the Moon, the engines must be started and stopped several times. Liquid engines do this by simply starting or stopping the flow of propellants into the combustion chamber.

Solid-propellant rockets are not as easy to control as liquid rockets. Once started, the propellants burn until they are gone. The burn rate of solid propellants is carefully planned in advance. The hollow core running the length of the propellants can be made into a star shape. At first, there is a very large surface available for burning, but as the points of the star burn away, the surface area is reduced. For a time, less of the propellant burns, and this reduces thrust.

In addition, a Velocity Trimming Module with liquid propellant thrusters are used as an additional stage to make necessary correction. For example, all long-range solid propellant missiles like Agni, MX have thrusters to make proportional correction, at the last phase of flight.

b. Stability and Control Systems

The rocket must also be stable in flight. A stable rocket is one that flies in a smooth, uniform direction. An unstable rocket flies along an erratic path, sometimes tumbling or changing direction. Unstable rockets are dangerous because it is not possible to predict where they will go. They may even turn upside down and suddenly head back directly to the launch pad.

Making a rocket stable requires some form of control system. It is first important to understand what makes a rocket stable or unstable. All matter, regardless of size, mass, or shape, has a point inside called the centre of mass (CM). The centre of mass is the exact spot where all of the mass of that object is perfectly balanced. The centre of mass is important in rocket flight because it is around this point that an unstable rocket tumbles. As a matter of fact, any object in flight tends to tumble.

DIFFERENT TYPES OF CONTROLS

In flight, spinning or tumbling takes place around one or more of three axes. They are called roll, pitch, and yaw. The point where all three of these axes intersect is the centre of mass, CM. For rocket flight, the pitch and yaw axes are more important because any movement in either of these two directions can cause the rocket to go off course. The roll axis is the not so important because movement along this axis will not affect the flight path. In fact, a rolling motion will help stabilise the rocket flight. Any variation in thrust from the engine or other perturbations while the rocket is in flight, may cause unstable motions in the pitch and yaw axes. This will cause the rocket to leave the planned course. To prevent this, a control system is needed. Different controls are shown in Figures 5.23 and 5.24.

(a)   (b)

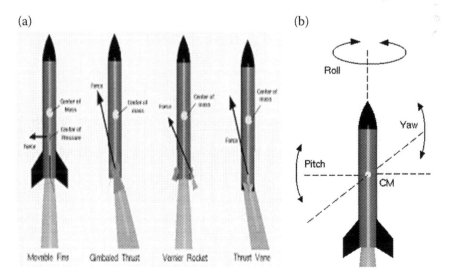

FIGURE 5.23   a) CM-centre of mass, b) Rotational axes; roll, pitch and yaw.

Courtesy: NASA.

(a)   (b)

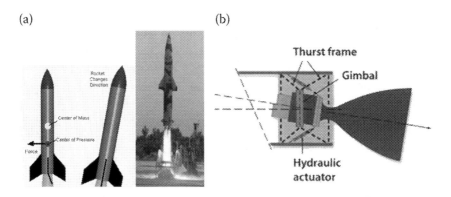

FIGURE 5.24   a) Movable fins, b) Gimbal control of engines.

Controls on rockets can either be active or passive. Passive controls are fixed devices that keep rockets stabilised by their very presence on the rocket's exterior. Active controls can be moved to steer the rocket to its flight path.

Stability is improved in rocket with clusters of lightweight fins mounted around the lower end near the nozzle, in different shapes and angles, and sometime with movable fin-tips, based on stability and control calculations. Fins could be made out of lightweight materials. They give rockets a dart like appearance. The large surface area of the fins easily keeps the centre of pressure behind the centre of mass, as we learnt in the last chapter for stability. But this design also produces more drag and limits the rocket's range. With the start of modern rocketry in the 20th century, new ways were sought to improve rocket stability and at the same time reduce overall rocket weight. The answer to this was the development of active controls.

Active control systems include jet vanes, movable fins, canards, gimbaled/flex nozzles, Vernier rockets, secondary injection thrust vectoring, fuel injection, and attitude-control rockets. Tilting fins and canards are quite similar to each other in appearance. The only real difference between them is their location on the rockets. Canards are mounted on the front end of the rocket while the tilting fins are at the rear. In flight, the fins and canards tilt like rudders to deflect the air flow and cause the rocket to change course. Motion sensors on the rocket detect unplanned directional changes, and corrections can be made by slight tilting of the fins and canards. The advantage of these two devices is size and weight. They are smaller and lighter and produce less drag than the large fins. Certain active control systems can eliminate fins and canards altogether. By tilting the angle at which the exhaust gas leaves the rocket engine, course changes can be made in flight. Several techniques can be used for changing exhaust direction.

Jet vanes are small finlike devices that are placed inside the exhaust of the rocket engine. Tilting the vanes deflects the exhaust, and by action-reaction the rocket responds by pointing the opposite way.

Controls in SLV-3: First stage with SITVC, Fin-tip (aerodynamic) control, second and third stages with bi-propellant and mono-propellant reaction control, respectively, and fourth stage with spin stabilisation.

	Stage 1	Stage 2	Stage 3	Stage 4
Pitch	SITVC	Engine Gimbal	Flex Nozzle	Engine Gimbal
Yaw	SITVC	Engine Gimbal	Flex Nozzle	Engine Gimbal
Roll	RCT and SITVC in 2 PSOMs	HRCM Hot Gas Reaction Control Motor	PS4 RCS	PS4 RCS

CONTROL FOR PSLV TYPE LV

Another method for changing the exhaust direction is to gimbal the nozzle. A gimbaled nozzle is one that is able to sway while exhaust gases are passing through it. By tilting the engine nozzle in the proper direction, the rocket responds by changing course. The other way is to use Vernier rockets mounted on the outside of the large engine. When needed, they fire, producing the desired course change.

SATELLITE ATTITUDE CONTROL

In space, only by spinning the rocket upper stage along the roll axis or active control the flight is stabilised before injecting the satellite. The most common kinds of active control used in satellites are attitude-control thrusters (Bi-propellant N_2O_4 + MMH or Mono-propellant Hydrazine). Clusters of small engines are mounted all around. By firing the right combination of these small thrusters, the satellite can be put in the right orbit and orientation, and its allotted station in the orbit. General cosed loop control is explained in Figure 5.25.

FLEX NOZZLE

Of all the mechanical deflection types, the flexible nozzles are the most efficient for solid rocket motor. They do not significantly reduce the thrust or the specific impulse and are weight-competitive with the other mechanical types. The flexible

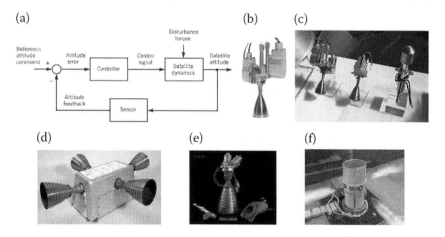

FIGURE 5.25 a) General closed-loop attitude control system, b) 10 N N_2O_4 + MMH thruster mono propellant hydrazine thrusters, c) LV high thrust RCS, d) Apollo reaction thrusters, e) Gemini RCS 100 N, and f) RCS testing (Northrop Grumman).

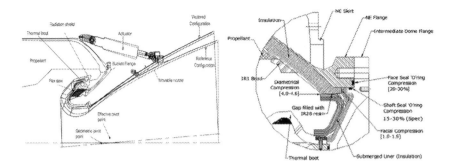

FIGURE 5.26 Flexible nozzle.

nozzle, shown in Figure 5.26 is a common type of TVC used with solid propellant motor. The moulded, multilayer bearing pack acts as a seal, a load transfer bearing, and a viscoelastic flexure. It uses the deformation of a stacked set of doubly curved elastomeric (rubbery) layers between spherical metal sheets to carry the loads and allow an angular deflection of the nozzle axis. The flexible seal nozzle has been used in launch vehicles and large strategic missiles.

At low temperature the elastomer becomes stiff, and the actuation torques increase substantially, requiring a much larger actuation system. There are double seals to prevent leaks of hot gas and various insulators to keep the structure below 100°C. It supports the weight of the engine and transmits the thrust force. The maximum angular motion is actually larger than the deflection angle during operation so as to allow for various tolerances and alignments. One version of this nozzle can deflect 4° maximum plus 0.5° for margin and another is rated at 7.5°. It has two electrically redundant electromechanical actuators using ball screws, two potentiometers for position indication, and one controller that provides both the power drive and the signal control electronics for each actuator. A variable-frequency, Pulse-Width-Modulated (PWM) electric motor drive is used to allow small size and low weight for the power and forces.

NAVIGATION AND GUIDANCE

The guidance system of a rocket includes very sophisticated sensors such as gyros and accelerometers, on-board computers, mission software and communication with control system. Guidance is the process of calculating the changes in position, velocity, altitude, and/or rotation rates of the rocket, which is required to follow a certain trajectory and/or altitude profile and help to implement necessary correction.

A guidance system has three major sub-sections: Inputs, Processing, and Outputs. The input section includes sensors, course data, radio and satellite links, and other information sources. The processing section, composed of one or more OBCs (On-Board Computer), integrates this data and determines correction necessary to maintain or achieve a proper heading. This is then fed to the controls, which can directly implement the system's course. The outputs may control trajectory by interacting with devices such as propulsion engine, thrusters, etc.

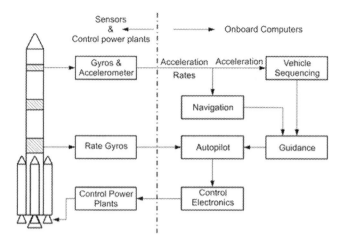

FIGURE 5.27 Navigation, guidance and control.

FUNCTION

Sensor data like those from compasses, GPS receivers, star trackers, inertial measurement units, altimeters, etc. are used as inputs for the navigation. The output of the navigation system, the navigation solution, is an input for the guidance system, among others like the environmental conditions (wind, water, temperature, etc.) and the vehicle's characteristics (i.e., mass, control system availability, control systems correlation to vector change, etc.). In general, the guidance system computes the instructions for the control system, which comprises the object's actuators (e.g., thrusters, reaction wheels, body flaps, etc.), which can manipulate the flight path and orientation of the object without direct or continuous human control, as explained in Figure 5.27.

INERTIAL GUIDANCE SYSTEM

Inertial guidance system continuously monitors the position, velocity, and acceleration of a rocket and thus provides navigational data for control of the trajectory without need for communicating with a base station. The basic components of an inertial guidance system are gyroscopes, accelerometers, and a computer. The gyroscopes provide fixed reference directions or turning rate measurements, and accelerometers measure changes in the velocity of the system. The computer processes information on changes in direction and acceleration and feeds its results to the vehicle's navigation system.

SINS (STABILISED PLATFORM INS)

There are two fundamentally different types of inertial navigation systems: gimbaling platform (SINS) and strap-down systems (SDINS). A typical gimbaling inertial navigation system uses three gyroscopes and three accelerometers. The three gimbal-mounted gyroscopes establish a frame of reference for the vehicle's roll (rotation about the axis running from the front to the rear of the vehicle), yaw (rotation about

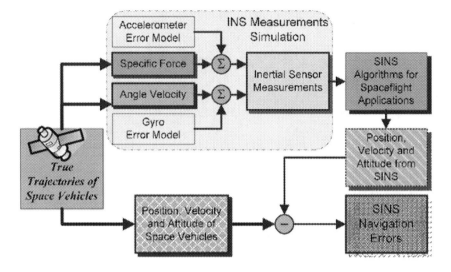

FIGURE 5.28   SINS (stabilised platform INS).

the axis running left to right), and pitch (rotation about the axis running top to bottom). The accelerometers measure velocity changes in each of these three directions. The computer performs two separate numerical integrations on the data it receives from the inertial guidance system. First, it integrates the acceleration data to get the current velocity of the vehicle; then, it integrates the computed velocity to determine the current position (Figure 5.28). This information is compared continuously to the desired (predetermined and programmed).

In a strap-down inertial navigation system (SDINS), the accelerometers are rigidly mounted parallel to the body axes of the vehicle. In this application, the gyroscopes do not provide a stable platform; they are instead used to sense the turning rates of the craft. Double numerical integration, combining the measured accelerations, and the instantaneous turning rates allows the computer to determine the craft's current velocity and position and to guide it along the desired trajectory.

In many modern inertial navigation systems, such as those used on commercial jetliners, rockets, and orbiting satellites, the turning rates are measured by ring laser gyroscopes (RLG) or by fibre-optic gyroscopes (FOG). Minute errors in the measuring capabilities of the accelerometers or in the balance of the gyroscopes can introduce large errors into the information that the inertial guidance system provides. These instruments must, therefore, be constructed and maintained to strict tolerances, carefully aligned, and reinitialised at frequent intervals using an independent navigation system such as the global positioning system (GPS). All types of sensors are explained in Figure 5.29 and accuracies in Figure 5.30.

In the case of the Prithvi missile, a higher level of target impact accuracy was obtained by injecting error modelling software in the conventional moderate accuracy sensors and data fusion from multiple GPS, by intelligent software developed indigenously. All accuracy values are established through extensive hardware-in-loop simulation (HILS) in 5-axis motion simulator, integrated with all sensors and control systems with 6 DOF flight trajectory.

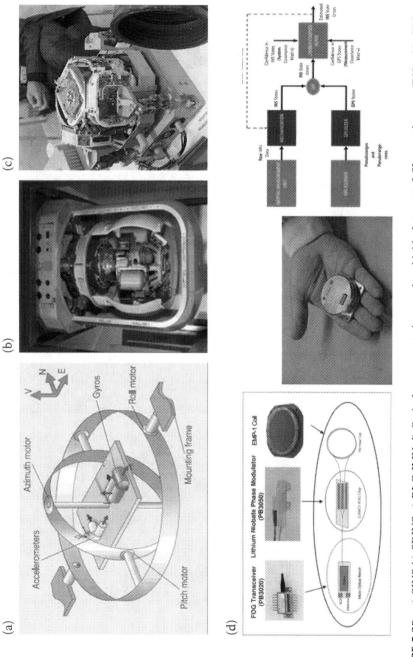

FIGURE 5.29 a) SINS b) SDINS c) RLG IMU d) Optical components in a modern highly **integrated fibre optic gyro**, INS with GPS.

Courtesy: Aerospace & Defence Technology, nedaero.com, blogs.nasa.gov.

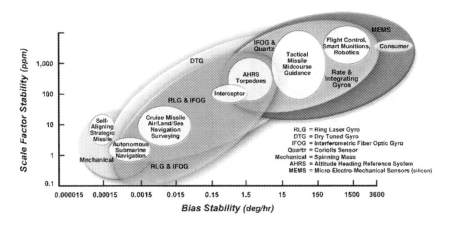

FIGURE 5.30   Sensors – Bias stability vs. scale factor stability.

GUIDANCE SENSORS ACCURACY AND APPLICATIONS

The accuracy and applications of various types of sensors and the bias stability vs. scale factor stability is given in Figure 5.30.

The typical equipment-bay shown in Figure 5.31, situated above third stage, houses INS, OBC, battery, and power distribution unit, 1553 bus-interfaces boxes, telemetry, telecommand, sensors, and controlling whole mission, until the injection of the satellite in the desired orbit with the required velocity.

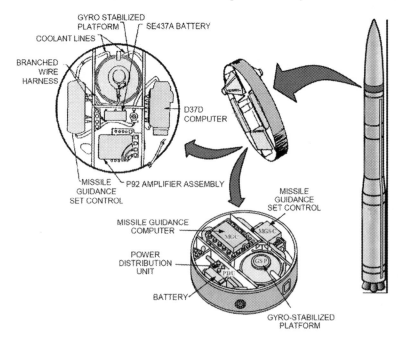

FIGURE 5.31   Guidance systems and its position in typical missile/launch system.

6 Rocket Design Methodology, Test and Evaluation, World Launch Sites

"Ultimately, education in its real sense is the pursuit of truth. It is an endless journey through knowledge and enlightenment."

- A. P. J. Abdul Kalam

LAUNCH VEHICLE DESIGN METHODOLOGY

Rockets used for injecting satellites in different desired orbits are called satellite launch vehicles. We studied the principle of a rocket, its subsystems, and components in earlier chapters. Design of a rocket is highly interactive, involving multiple disciplines and technologies. In this chapter, let us study the design methodology.

Let's say we decide to build a new rocket. There are different types of rockets – a launch vehicle to orbit a satellite in a desired altitude for space application or scientific experiment, or injecting a spacecraft for space exploration, or injecting a payload module carrying humans to Mars, or a long-range surface-to-surface guided missile with an explosive warhead as payload. These rockets are designed for different purposes.

The first step is to know specific purpose – say, to orbit a satellite in low Earth orbit. Secondly, the mission parameters such as satellite weight, orbit – circular or elliptical (apogee, perigee), orbital inclination with respect to equator, expected lifetime and orbital velocity – must be established. Thirdly, a rough configuration of the rocket, number of stages, and type of propulsion with a 2D trajectory run can be carried out. There could be many configuration options with take-off weight, payload, velocity, and acceleration profiles at various altitudes and loads. Wind tunnel tests of different configurations will provide critical aerodynamic coefficients. Preliminary Computational Fluid Dynamic (CFD) studies will lead to better configuration.

DOI: 10.1201/9781003323396-6

Once a particular configuration is chosen, preliminary design of the rocket is carried out detailing propulsion, structure, control, guidance, and satellite along with aerodynamics, accurate model testing, CFD, aero elastic studies, wind conditions, and trajectory optimisation with 6 DOF (degree of freedom) software, which gives an integrated performance. The details are put through a review, called PDR (preliminary design review) by experts. Choice of materials, manufacturing methodologies, testing requirements, quality assurance plans, infrastructure and cost, and development strategies are also considered during PDR. Development of a rocket starts with appropriate Project Management Organisation, contracting, supply chain management, and milestone plans, linked with availability of funds.

We can consider a satellite launch vehicle, SLV-3, and interplanetary mission to Mars, as examples.

SLV-3

Let us take the example of SLV-3, the first Satellite Launch Vehicle of India. The SLV-3 is designed to inject Rohini satellite of 40 kg in a low earth orbit, which needs a minimum velocity of 7.88 km/s. The satellite must be injected at an altitude of 300 km and at an inclination of 45° to the equator, when launched from SHAR.

LAUNCH VEHICLE DESIGN – A SIMPLIFIED FORM

The design process includes (Figure 6.1):

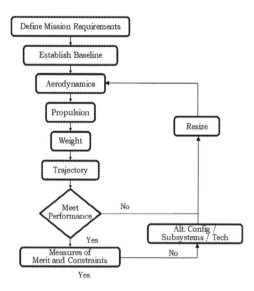

FIGURE 6.1 Flow chart of typical design process.

- specifying the mission requirements (e.g., satellite payloads, size, mass, desired orbit, environmental constraints, on-orbit operations).
- selecting a vehicle approach (baseline configuration – staging, velocity build-up, propulsion modules).
- selecting technologies (e.g., structural materials, control system, guidance, avionics,).
- 2D trajectory and firming of technology choice, performance, constraints creating a physical layout and surface geometry of the rocket that will contain subsystems, and satellite.
- estimating the ascent and entry aerodynamics (subsonic, transonic, supersonic, hypersonic).
- optimising trajectory (6D) with control-structure – propulsion and the resulting flight environments.
- reviewing structural, controls, heating, radiation, and propulsion analyses based on the flight environment.
- estimating the vehicle weights, dimensions, and centre of gravity based on layout, flight environment, and confirming technology selection and firm design.
- analysing operations, maintainability, hardware/software requirements, reliability, and safety based on operational scenario, vehicle configuration, and technologies.
- estimating development time and costs (e.g., design, development, test, evaluation).
- calculating performance and programmatic evaluation criteria (metrics) used to compare alternatives.
- optimise and modify the overall system to better meet mission requirements and design objectives.
- conceptual review and approval of mission and design.
- detailed design – subsystem wise, wind tunnel testing, CFD analysis, preliminary design review (PDR).

This leads to preparation of a detailed project report, project approval, and funding. Development starts for subsystems by technology groups, followed by tests and evaluation to extreme flight conditions, quality assurance, configuration control, and systematic reviews including Critical Design Review (CDR), Flight Readiness Review (FRR), and Mission Readiness Review (MRR).

ROCKET SYSTEMS

The main energy source is the four solid rocket motors. The solid rocket motors with their control system, heat shield, stage separation systems, guidance, on-board computer, satellite and its attitude control system, and instrumentation are sequenced, and the flight path is controlled, based on the attitude reference input processed and fed to the control system for the required control force.

The first stage of SLV-3 has been designed for the required ballistic performance, with 54 T thrust and structurally integrated, including viscoelastic analysis. The other three stage rocket motors thrust rating are about 29T, 9T, and 2T. The heat shield protects the fourth stage rocket motor and the satellite from aerodynamic heating.

SLV-3 has 44 major sub-systems with 100,000 components of mechanical, electrical, electromechanical, and chemical engineering, and also mission software and interfaces. If even one component fails to function to the requirements, it results in the entire mission's failure.

CHECK-OUT

The total design, apart from individual components and systems, considers the interface requirements between the stages and systems. From the drawing board to launch, it is essential for launch vehicle designers to conceive the functions of each component relating to sub-system, system performance, and interfacing ground installations like the automatic checkout system. From launch vehicle design and development to the first successful launch, 54 static tests of rocket motors have been conducted at various environmental conditions, and their thrust time performance have been confirmed with prediction.

CONTROL SYSTEM

For effective control of SLV-3, Fin Tip Control (FTC), and Thrust Vector Control (TVC) for the first stage, Bi-propellant Reaction Control System (RCS) for second stage, Mono-propellant RCS for third stage, and spin-stabilization for the fourth stage were employed. Thousands of thrusters/components have been evaluated with various instrumentation packages, including equipment bay. Also, they have been environmentally tested for actual performance and qualified through seven sounding rocket flights.

GUIDANCE

SLV3 employed open loop guidance with stored pitch programme to steer the vehicle in flight along the pre-determined trajectory, with the mission software loaded in the vehicle attitude programmer. E-80 IMU was used for guidance. The LV was heavily instrumented for measuring 600 parameters.

TRAJECTORY DESIGN

The launch vehicle during its flight passes through various flight regimes like subsonic, transonic, supersonic, and hypersonic, finally achieving the required injection velocity. Aerodynamic, thermal, and structural design for launch vehicle structures had been carried out for both static and dynamic loads, which the launch vehicle would experience during its ascent phase.

SEQUENCE OF TASKS BEFORE FLIGHT

The sequence of tasks to be carried out before the flight trial is explained in Figure 6.2. Propulsion, airframe, guidance & control, power supply, avionics, and

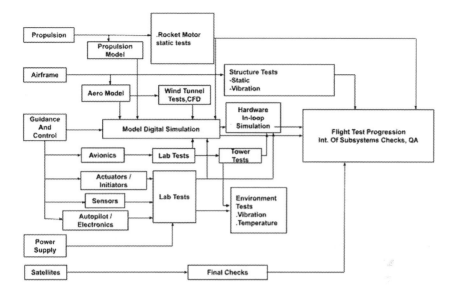

FIGURE 6.2   Sequence of tasks – Subsystems to flight.

satellite go through systematic model testing, simulations, and qualification tests before assembly and flight testing, as explained in the sequence.

DEVELOPMENT OF DESIGN AND OPERATIONS SOFTWARE FOR SLV-3

The automatic checkout system consists of computerized ground telemetry, tele-command, payload, control and guidance, and power systems. When the sub-systems are realized at the same time, the automatic checkout system is also developed. The system is then interfaced with on-board systems and various systems functioning is carried out using the checkout system. The launch vehicle is one of the elements of this total mission that must be integrated with the launch complex, downrange telemetry station, and other satellite tracking network and safety systems, as explained in Figure 6.3.

Given a mission, the configuration design is carried out, linking the aero-dynamic, structural, propulsion, control, and guidance designs. Apart from vehicle performance, the launch vehicle design considers the launch azimuth based on instantaneous impact points and fly-over countries criteria, launch window based on the satellite injection time, and the surface and in-flight wind conditions. In addition, during the vehicle development phase, to study the performance of flight, 600 parameters had to be telemetered. The post-flight analysis methodology using these telemetered data in real-time was evolved. For design alone, the total computer time used was about 30,000 hours using various computing systems like the IBM 360/44, PDP 11/34, IRIS-55, EAI 682/ PACE-100, etc.

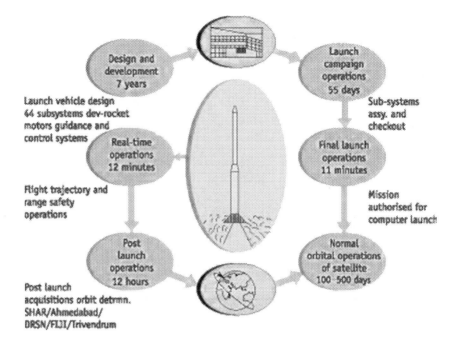

FIGURE 6.3   Design and development cycle – Development to launch, satellite in orbit.

TRAJECTORY DESIGN OF ISRO'S MARS MISSION

Hohmann transfer orbit between Earth and Mars, as used by the ISRO PSLV-C25

GEO CENTRIC PHASE

The spacecraft (Figure 6.4a) was injected into an Elliptic Parking Orbit by PSLV-C-25 (Figure 6.4b) on 25 November 2013. With six main engines burnt, the spacecraft was gradually manoeuvred into a departure hyperbolic trajectory with which it escaped from the Earth's Sphere of Influence (SOI) with Earth's orbital velocity + V boost. The orbiter crossed SOI of Earth at 918,347 km from the surface of the Earth, beyond which the perturbing force on the orbiter was due to the Sun. One primary concern was how to get the spacecraft to Mars, on the least amount of fuel. ISRO used a **Hohmann Transfer Orbit** – or a Minimum Energy Transfer Orbit – to send a spacecraft from Earth to Mars with the least amount of fuel possible.

HELIO-CENTRIC PHASE

The spacecraft travelled in a flight path is roughly one half of an ellipse around the Sun. This trajectory became possible with certain allowances when the

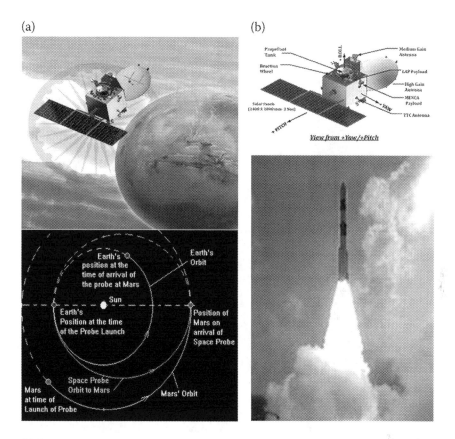

FIGURE 6.4   a) Mars spacecraft b) PSLV-C-25.

relative position of Earth, Mars, and the Sun was at an angle of approximately 44°. Such an arrangement recurs periodically at intervals of about 780 days. Minimum energy opportunities for Earth-Mars occur in November 2013, January 2016, May 2018, etc.

MARTIAN PHASE

The spacecraft arrived at the Mars SOI (around 573,473 km from the surface of Mars) in a hyperbolic trajectory. At the time, the spacecraft reached the closest approach to Mars (Periapsis), it was captured into planned orbit around Mars by imparting ΔV retro, which is called the Mars Orbit Insertion (MOI) manoeuvre on 24 September 2014. as in Figure 6.5.

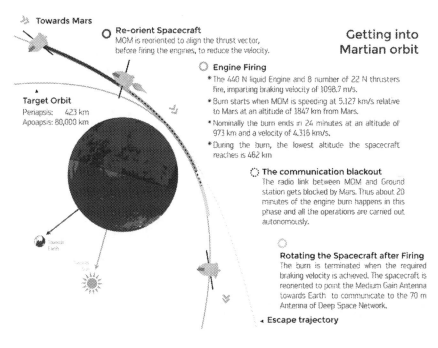

FIGURE 6.5   Getting into Martian orbit.

HUMAN MISSION TO MARS (INTER PLANETARY TRAVEL)

First, the mission requirements must be spelt out. It is an interplanetary travel from Earth to Mars with three astronauts on-board. It will be highly exciting like the Apollo-11 with three astronauts to the Moon, two of them landing on the surface of the Moon in July 1969. But this mission will be more complex, as we saw the Perseverance rover had to go through before landing. Considering the distance to Mars and the duration of travel for astronauts, it will be the most ambitious task ever undertaken in spaceflight history, to land on another planet.

We have a history of many missions to Mars, either orbiting satellites or landing of rovers by the USA and China. India's MOMs orbiter mission was a great success. The USA and China are on the lead with successful rover missions. Space-X is also in the race. Who will be the first to land humans on the surface of Mars? Will it be known soon?

The mission may need a heavy payload of about 150 metric tons lifting to LEO with a large volume lander/orbiter for a long duration support to the astronauts.

Many configurations are possible. Choice of propulsion, trajectory design, environmental conditions during decent and ascent phases crossing Mars atmosphere and life support system will be crucial. Experience of long duration stay in ISS by astronauts could be helpful, but Mars travel will not be benign. Several rounds of discussions with different experts and simulations will ensure near success, and the reviews will close the rocket design. The challenge is designing the best rocket optimally chosen for the mission.

NASA when decided on SLS four years before, went through vigorous reviews and finalised a modular configuration for several future missions, including missions to the Moon and Mars with humans on board.

A selection of just a few of the major designs considered in early SLS studies. Three stood out – a Saturn-like large, multi-stage rocket, using the kerosene fuel that powered the Moon rocket instead of the shuttle's liquid hydrogen; a rocket built from components based on current smaller rockets, taking advantage of industry successes; and a design that would be an evolutionary step from systems used on the space shuttle. Any choice will be based on minimum additional resources, safety, and sustainability.

WIND TUNNEL TESTING AND CFD

Wind tunnels are tube-shaped facilities that allow moving air over an instrumented vehicle model if it were flying, at different angle of attacks as per the flight regime.

Computational fluid dynamics (CFD) uses computers to determine the flow. Because of the rapid growth of computing capacity and the development of efficient computation schemes, CFD has become popular in many fields of fluid dynamics. However, the separate flows around a structure are complicated, and it generally is difficult to obtain a quantitative prediction of the aerodynamic force and response using CFD. To overcome this problem, extensive studies on utilising CFD in the field of bridge aerodynamics have been and are currently being performed. For instance, a streamline box girder was analysed using an elaborate numerical model where even railings were reproduced, and the obtained steady and unsteady aerodynamic coefficients agreed well with experimental results. In another example, a numerically less demanding model was used to obtain the coefficients, which also agreed reasonably well with experimental results. Although it may be still difficult to use CFD for the final estimation of the bridge response to wind, the results by CFD are already used at the first stage of wind-resistant design. The wind tunnel tests include measurements of six component forces, pressure, and flow visualisation. CFD results matched with results of WT tests.

THE FINAL CHOICE IS SPACE LAUNCH SYSTEM (SLS) FOR MOON, MARS

NASA's Space Launch System, is a heavy lift launch vehicle with unprecedented power and capabilities and is designed for missions, including humans to the Moon and Mars and robotic scientific spacecraft to the Moon, Mars, Saturn, and Jupiter. The first mission will be Artemis-1.

Every SLS configuration uses the core stage with four RS-25 engines. The first SLS vehicle, called Block 1, can send more than 27 metric tons (t) to orbits beyond the Moon. It will be powered by twin five-segment solid rocket boosters and four RS-25 liquid propellant engines. After reaching space, the Interim Cryogenic Propulsion Stage (ICPS) sends Orion on to the Moon. The first three Artemis missions will use a Block 1 with an ICPS.

Block 1B crew vehicle will use a new, more powerful Exploration Upper Stage (EUS) to enable more ambitious missions. The Block 1B vehicle can, in a single launch, carry the Orion crew vehicle along with large cargoes for exploration systems needed to support a sustained presence on the Moon. The Block 1B crew vehicle can send 38 t to deep space including Orion and its crew.

The next SLS configuration, Block 2, will provide 9.5 million lbs. of thrust and will be the workhorse vehicle for sending cargo to the Moon, Mars, and other deep space destinations. SLS Block 2 will be designed to lift more than 46 t to deep space.

VAN ALLEN RADIATION BELTS

A Van Allen radiation belt is a zone of energetic charged particles, most of which originate from the solar wind, that are captured by and held around a planet by that planet's magnetosphere and some from cosmic rays. Earth has two such belts. The belts are named after James Van Allen, who is credited with their discovery. Earth's two main belts extend from an altitude of about 640 to 58,000 km above the surface, in which region radiation levels vary particles. By trapping the solar wind, the magnetic field deflects those energetic particles and protects the atmosphere from destruction.

The belts are in the inner region of Earth's magnetic field. The belts trap energetic electrons and protons. Other nuclei, such as alpha particles, are less prevalent. The belts endanger satellites, which must have their sensitive components protected with adequate shielding if they spend significant time near that zone. So, satellites passing through are radiation hardened.

WORLD LAUNCH SITES

Launch sites also influence the launch vehicle capability to deliver the satellite in particular orbit, especially due to range safety considerations, over-flying land mass involving international regulations and the latitude from equator (for geo orbit inclination correction needing additional fuel in satellite).

The ideal launch site is location at equator with due east launch for GEO mission, due south, or north launch for polar mission. Satellite weight is saved, better cost/kg. Kourou of French Guiana is the best station of all. Sriharikota has range safety constraints leading to loss of payload wight. Polar launces from SHAR need dog-leg manoeuvre to avoid overflying Sri Lanka, reducing vehicle performance. Due east GEO missions from SHAR need overflying Malaysia. Russia has communication satellites in inclined orbits, not 24-hour service, needing more satellites. Suitably launch vehicle design will change. Location of launch sites in the world are given in Figure 6.6 – launch stations of India (Figure 6.7a), Japan (Figure 6.7b), Russia (Figure 6.7c), Europe (Figure 6.7d), the USA (Figure 6.7e), and China (Figure 6.7f).

Orbital launch sites by payload class

☐ Government
⊡ Commercial
■ Government/commercial

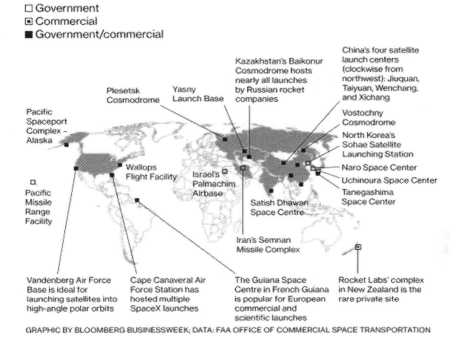

GRAPHIC BY BLOOMBERG BUSINESSWEEK; DATA: FAA OFFICE OF COMMERCIAL SPACE TRANSPORTATION

FIGURE 6.6 Launch sites of space faring countries.

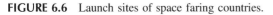

FIGURE 6.7a India's space port.

FIGURE 6.7b Japan launch station.

FIGURE 6.7c Russia launch station.

FIGURE 6.7d Europe's spaceport (CSG-Kourou, French Guiana) Ideal launch site for maximum efficiency for both Geo and Polar missions.

FIGURE 6.7e JF Kennedy space centre at Florida, USA.

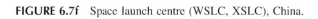

FIGURE 6.7f   Space launch centre (WSLC, XSLC), China.

7 Satellites, Orbits and Missions

SATELLITE

Earth is a natural satellite, and it orbits around the Sun. Likewise, the Moon is a satellite because it orbits around the Earth. Usually, the word "satellite" refers to a machine that is launched into space and moves around Earth or another body in space.

SPUTNIK-1 AND THE DAWN OF THE SPACE AGE

On 4 October 1957, the Soviet Union successfully launched artificial satellite Sputnik I (Figure 7.1). The world's first artificial satellite was about the size of a ball (58 cm diameter), weighed 83.6 kg. It took about 98 minutes to orbit Earth at 8 km/s speed at 577 km altitude and made 1440 orbits.

(a)   (b)

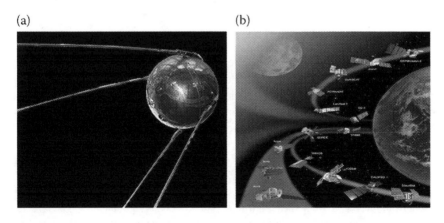

FIGURE 7.1 (a) SPUTNIK-1 (1957) (b) NASA satellites now in orbits.

Source: https://serc.carleton.edu/.

Today, thousands of artificial, or man-made, satellites orbit around the Earth, launched by many countries. Some satellites are used mainly for communications, such as beaming TV signals and phone calls around the world. A group of satellites make up the Global Positioning Systems, or GPS. Some take pictures of the planet that help meteorologists predict weather and track hurricanes. Some take pictures of other planets, or faraway galaxies. These pictures help scientists better understand the solar system and universe. Typical satellites of NASA are shown in Figure 7.1(b)

DOI: 10.1201/9781003323396-7   **129**

IMPORTANCE OF SATELLITES

Satellites can see large areas of Earth at one time. This ability means satellites can collect more data, more quickly, than instruments on the ground. Satellites also can see into space better than telescopes at Earth's surface. That's because satellites fly above the clouds, dust and molecules in the atmosphere that can block the view from ground level.

Before satellites, TV signals didn't go very far. TV signals only travel in straight lines. So, they would quickly trail off into space instead of following Earth's curve. Sometimes mountains or tall buildings would block them. Phone calls to faraway places were also a problem. Setting up telephone wires over long distances or underwater is difficult and costs a lot. With satellites, TV signals and phone calls are sent upward to a satellite. Then, almost instantly, the satellite can send them back down to different locations on Earth.

Satellites come in many shapes and sizes. But most have at least two parts in common – an antenna and a power source. The antenna sends and receives information, often to and from Earth. The power source can be a solar panel or battery. Solar panels make power by turning sunlight into electricity.

Many Earth observation satellites carry cameras and scientific sensors. Sometimes these instruments point toward Earth to gather information about its land, air, and water. Other times they face toward space to collect data from the solar system and universe.

ORBIT

Our understanding of orbits dates back to Johannes Kepler in the 17th century. In general sense, an orbit is a curved path that one object in space takes around another one. The object in an orbit is called a satellite. A satellite can be natural, like Earth or the Moon or man-made such as a communication satellite, orbiting around the Earth. If the spacecraft comes near a large body in space, the gravity of that body will unbalance the forces and curve the path of the spacecraft.

The combination of a satellite's forward motion and the pull of gravity of the planet bend the satellite's path into an orbit. This happens when a satellite is sent by a rocket on a path that is tangent to the planned orbit about a planet. The unbalanced gravitational force causes the satellite's path to change to an arc. The arc is a combination of the satellite's fall inward toward the planet's centre and its forward motion. When these two motions are just right, the shape of the satellite's path matches the shape of the body it is travelling around. Consequently, an orbit is produced. Since the gravitational force changes with height above a planet, each altitude has its own unique velocity that results in a circular orbit.

Obviously, controlling velocity is extremely important for maintaining the circular orbit of the spacecraft. Unless another unbalanced force, such as friction with gas molecules in orbit or the firing of a rocket engine in the opposite direction, slows down the spacecraft, it will orbit the planet forever. The time it takes a satellite to make one full orbit is called its period. For example, Earth has an orbital period of one year. The inclination is the angle the **orbital plane** makes when compared with Earth's equator.

ORBITAL VELOCITY OF SATELLITE

The centripetal force required to keep the satellite in circular orbit is

$$F = \frac{mv_o^2}{r} = \frac{mv_o^2}{R + hr}$$

The gravitational force between the Earth and the satellite is

$$F = \frac{GMm}{r^2} = \frac{GMm}{(R + h)^2}$$

For the stable orbital motion,

$$\frac{mv_o^2}{R + h} = \frac{GMm}{(R + h)^2}$$

$$v_o = \sqrt{\frac{GM}{R + h}}$$

If the satellite is at a height of few hundred kilometres (say 200 km), (R + h) could be replaced by R.

$\therefore$ Orbital velocity, $v_o = \sqrt{gR}$

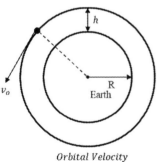

Orbital Velocity   **FIGURE 7.2**   Orbital velocity.

MINIMUM ORBIT – FIRST COSMIC VELOCITY

If the orbit is close to the Earth. H is negligible compared to R

$$\left(v = \sqrt{\frac{GM}{R}}\right) = \sqrt{Rg}$$

This orbit is called minimum orbit. The velocity corresponding to minimum orbit is called first cosmic velocity.

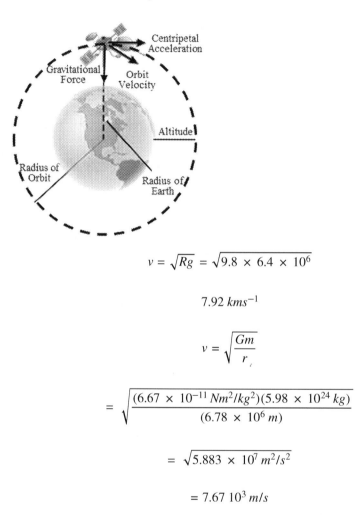

$$v = \sqrt{Rg} = \sqrt{9.8 \times 6.4 \times 10^6}$$

$$7.92 \; kms^{-1}$$

$$v = \sqrt{\frac{Gm}{r_e}}$$

$$= \sqrt{\frac{(6.67 \times 10^{-11} \, Nm^2/kg^2)(5.98 \times 10^{24} \, kg)}{(6.78 \times 10^6 \, m)}}$$

$$= \sqrt{5.883 \times 10^7 \, m^2/s^2}$$

$$= 7.67 \; 10^3 \, m/s$$

SATELLITE ORBITS

LOW EARTH ORBIT (LEO)

LEO is typically a circular orbit about 400–900 km above the Earth's surface and, correspondingly, has a much shorter period (time to revolve around the earth) of about 90 minutes. Because of their low altitude, these satellites are only visible from within a small area (about 1000 km radius) beneath the satellite as it passes overhead. In addition, satellites in low Earth orbit change their position relative to the ground position quickly. So even for local applications, a large number of satellites are needed if the mission requires uninterrupted connectivity. For this reason, LEO satellites are often part of a group of satellites working in concert otherwise known as a satellite constellation. Low Earth orbiting satellites are less expensive to launch into orbit than geostationary satellites and, due to proximity to the ground, do not require high signal strength.

Medium Earth Orbit (MEO)

MEO is the region of space around the Earth above low Earth orbit and below geostationary orbit. The most common use for satellites in this region is for navigation, such as the GPS (with an altitude of 20,200 km), GLONASS (with an altitude of 19,100 km) and Galileo (with an altitude of 23,222 km) constellations. Communications satellites that cover the North and South Pole are also put in MEO. The orbital periods of MEO satellites range from 2 to 24 hours. Telstar, one of the first and most famous experimental satellites, orbits in MEO. Navigation satellites, as a constellation in MEO, provide coverage across large parts of the world all at once.

Polar Orbit

A polar orbit is one in which a satellite orbits around the Earth between pole to pole, in a north-south direction, at 500–800 km. It has an inclination of near 90° to the equator. As Earth spins underneath, these satellites can scan the entire globe, one strip at a time, at a different longitude on each of its orbits due to the fact of the Earth's rotation. Polar orbits are often used for Earth-mapping, Earth observation, capturing the Earth as time passes from one point, reconnaissance satellites, as well as for some weather satellites.

Near-polar orbiting satellites commonly choose a Sun-synchronous orbit, meaning that each successive orbital pass occurs at the same local time of day. This can be particularly important for applications such as remote sensing atmospheric temperature, where the most important thing to see may well be *changes* over time. The orbital period is normally about 90–100 minutes.

Geo Stationary Orbit (Arthur C. Clarke Orbit)

A satellite in a geostationary orbit appears to be in a fixed position to an Earth-based observer. A geostationary satellite revolves around the Earth at a constant speed once per day over the equator. The geostationary orbit is useful for communication applications because ground-based antennas, directed toward the satellite can operate effectively without the need for expensive equipment to track the satellite's motion.

A geostationary orbit, also referred to as a geosynchronous equatorial orbit (GEO), is a circular geosynchronous orbit 35,786 km in altitude above Earth's equator (42,164 km in radius from Earth's centre). A geostationary satellite travels from west to east over the equator. It moves in the same direction and at the same rate Earth is spinning. From Earth, a geostationary satellite looks like it is standing still since it is always above the same location.

The launch vehicle injects the satellite in a high elliptical orbit called Geo Transfer Orbit (GTO) and then the apogee kick velocity is given to the satellite to go in GEO circular orbit over equator, with zero-degree inclination.

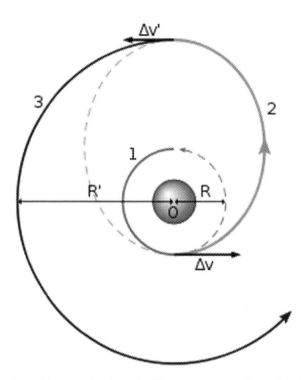

In orbital mechanics, the **Hohmann transfer orbit** is an elliptical orbit used to transfer between two circular orbits of different radii around a central body in the same plane. The Hohmann transfer often uses the lowest possible amount of propellant in travelling between these orbits, but bi-elliptic transfers can use less in some cases. The orbital manoeuvre to perform the Hohmann transfer uses two engine impulses, one to move a spacecraft onto the transfer orbit and a second to move off it. This manoeuvre was named after Walter Hohmann, the German scientist who published a description of it in his 1925 book Die Erreichbarkeit der Himmelskörper (The Attainability of Celestial Bodies). Hohmann transfer orbit, labelled 2, from an orbit (1) to a higher orbit (3)

The elliptic transfer orbits between different bodies (planets, moons, etc.) are often referred to as Hohmann transfer orbits. When used for travelling between celestial bodies, a Hohmann transfer orbit requires that the starting and destination points be at particular locations in their orbits relative to each other. Space missions using a Hohmann transfer must wait for this required alignment to occur, which opens a so-called launch window. For a space mission between Earth and Mars, for example, these launch windows occur every 26 months. A Hohmann transfer orbit also determines a fixed time required to travel between the starting and destination points; for an Earth-Mars journey this travel time is about nine months. When transfer is performed between orbits close to celestial bodies with significant gravitation, much less delta-v is usually required, as the Oberth effect may be employed for the burns.

They are also often used for these situations, but low-energy transfers which take into account the thrust limitations of real engines, and take advantage of the gravity wells of both planets can be more fuel efficient.

TABLE 7.1
Tangential Velocities at Altitude

Orbit	Centre-to-centre distance	Altitude above the Earth's surface	Speed	Orbital period	Specific orbital energy
Earth's own rotation at surface (for comparison – not an orbit)	6378 km	0 km	465.1 m/s (1674 km/h or 1040 mph)	23 hours 56 minutes 4.09 seconds	−62.6 MJ/kg
Orbiting at Earth's surface (equator) theoretical	6378 km	0 km	7.9 km/s (28,440 km/h or 17,672 mph)	1 hour 24 minutes 18 seconds	−31.2 MJ/kg
Low Earth orbit	6600–8400 km	200–2000 km	• Circular orbit: 6.9–7.8 km/s (24,840–28,080 km/h or 14,430–17,450 mph), respectively • Elliptic orbit: 6.5–8.2 km/s, respectively	1 hour 29 minutes–2 hours 8 minutes	−29.8 MJ/kg
Molinia orbit	6900–46,300 km	500–39,900 km	1.5–10.0 km/s (5400–36,000 km/h or 3335–22,370 mph), respectively	11 hours 58 minutes	−4.7 MJ/kg
Geostationary	42,000 km	35,786 km	3.1 km/s (11,600 km/h or 6935 mph)	23 hours 56 minutes 4.09 seconds	−4.6 MJ/kg
Orbit of the Moon	363,000–406,000 km	357,000–399,000 km	0.97–1.08 km/s (3492–3888 km/h or 2170–2416 mph), respectively	27.27 days	−0.5 MJ/kg

TABLE 7.2

First Launch by Country – Orbiting Satellite in LEO

Order	Country	Date of first launch	Rocket	Satellite
1	Soviet Union	4 October 1957	Sputnik	Sputnik 1
2	United States	1 February 1958	Juno I	Explorer 1
3	France	26 November 1965	Diamant-A	Astérix
4	Japan	11 February 1970	Lambda-4S	Ohsumi
5	China	24 April 1970	Long March 1	Dong Fang Hong I
6	UK	28 October 1971	Black Arrow	Prospero
7	India	18 July 1980	SLV-3	Rohini D1
8	Israel	19 September 1988	Shavit	Ofeq 1
9	Iran	2 February 2009	Safir-1	Omid
10	North Korea	12 December 2012	Unha-3	Kwangmyŏngsŏng-3 Unit 2
11	South Korea	30 January 2013	Naro-1	STSAT-2C
12	New Zealand	12 November 2018	Electron	CubeSat

VELOCITY REQUIREMENT FOR ORBITS AND ESCAPE FROM EARTH

Escape velocity is the speed at which the sum of an object's kinetic energy and its gravitational potential energy is equal to zero for an object, which has achieved escape velocity is neither on the surface, nor in a closed orbit (of any radius). The escape velocity from Earth's surface is about 11,186 m/s.

For a spherically symmetric, massive body such as a star, or planet, the escape velocity for that body, at a given distance, is calculated by the formula

$$K.E. = W$$

Or,

$$\frac{1}{2}mv_{e^2} = \frac{GMm}{R}$$

Where,

G is the universal gravitational constant ($G = 6.67 \times 10^{-11} \, m^3 kg^{-1} s^{-2}$)

M is mass of the body to be escaped from, and

R *is* the distance from the centre of mass of the body to the object

$$v_e = \sqrt{\frac{2GM}{R}}$$

$$v_{esc,} = \sqrt{\frac{2(6.67 \times 10^{-11})(6 \times 10^{24})}{(6378 \times 10^3)}}$$

$$= 11.2 \, km \, s^{-1}$$

SATELLITES IN ORBIT

Since Sputnik 1, about 8900 satellites from more than 40 countries have been launched. According to a 2018 estimate, 5000 remained in the orbit. Of those, about 1900 were operational, while the rest have exceeded their useful lives and become space debris. Approximately 63% of operational satellites are in low Earth orbit, 6% are in medium-Earth orbit (at 20,000 km), 29% are in geostationary orbit (at 35,786 km), and the remaining 2% are in various elliptical orbits. In terms of countries with the most satellites, the USA is the first with 1897 satellites, China is second with 412, and Russia is third with 176.

A few large space stations, including the International Space Station, have been launched in parts and assembled in orbit. Over a dozen space probes have been placed into orbit around other bodies and become artificial satellites of the Moon, Mercury, Venus, Mars, Jupiter, Saturn, a few asteroids, a comet, and the Sun.

MISSION PLANNING

A satellite mission involves a launch segment with a suitable launch vehicle, mission management at launch complex, command, control, communication, computer centres, orbit correction plans execution of the space segment linked to Mission Control, and utilisation of data received.

A launch vehicle is a rocket that places a satellite into orbit. Usually, it lifts off from a launch pad on land. Some are launched at sea from a submarine or a mobile maritime platform, or aboard a plane.

A typical trajectory of a communication satellite by ISRO is shown in Figure 7.3, giving the sequence of events during launch, involving an elaborate preparation of launch vehicle, satellite, mission control, range operations, real-time data management, etc.

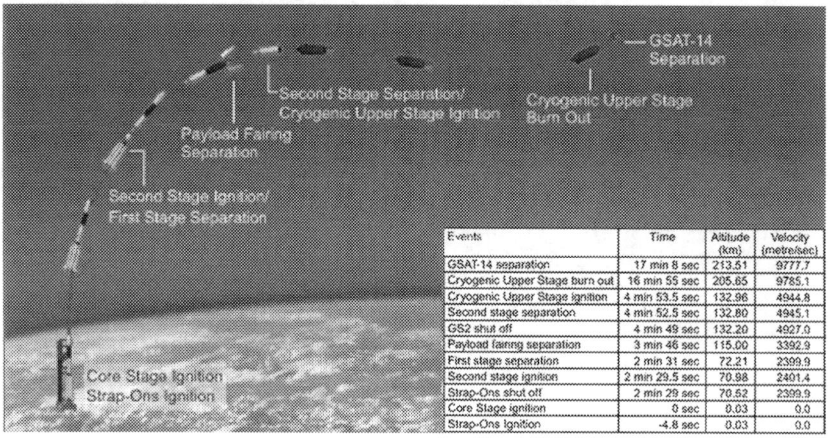

Events	Time	Altitude (km)	Velocity (metre/sec)
GSAT-14 separation	17 min 8 sec	213.51	9777.7
Cryogenic Upper Stage burn out	16 min 55 sec	205.65	9785.1
Cryogenic Upper Stage ignition	4 min 53.5 sec	132.96	4944.8
Second stage separation	4 min 52.5 sec	132.80	4945.1
GS2 shut off	4 min 49 sec	132.20	4927.0
Payload fairing separation	3 min 46 sec	115.00	3392.9
First stage separation	2 min 31 sec	72.21	2399.9
Second stage ignition	2 min 29.5 sec	70.98	2401.4
Strap-Ons shut off	2 min 29 sec	70.52	2399.9
Core Stage ignition	0 sec	0.03	0.0
Strap-Ons Ignition	-4.8 sec	0.03	0.0

FIGURE 7.3 Flight trajectory of GSLV-D5.

Courtesy: ISRO.

SATELLITE APPLICATIONS

One of the foremost advantages of space technology is that it can provide a change in the life of the common man. With its reach and potential benefits, people from all over the country are being benefited by space applications. In the past few decades, satellite-based communication, remote sensing technologies, and navigation have provided services related to education, healthcare weather, land and water resource management, mitigation of impact of natural disasters, route map and location, etc.

Out of the 2666 operational satellites circling the globe in April 2020, 1007 were for communication services. In addition, 446 are used for observing the Earth and 97 for navigation/ GPS purposes.

More than half of Earth's operational satellites are launched for commercial purposes. About 61% of those provide communications, including everything from satellite TV and Internet of Things (IoT) connectivity to the global Internet.

Communication satellites in Geo Orbit are a vital part of our infrastructure, helping us for day-to-day tasks.

These satellites are injected in geo-transfer orbit by launch vehicles and later satellites themselves give necessary velocity input to reach GEO. The applications include communication, broadcasting, weather prediction, disaster warning, and military needs with availability of 24 hours in a day. These communication satellites provide valuable support for national development. A typical application of INSAT communication satellite is given in Figure 7.4.

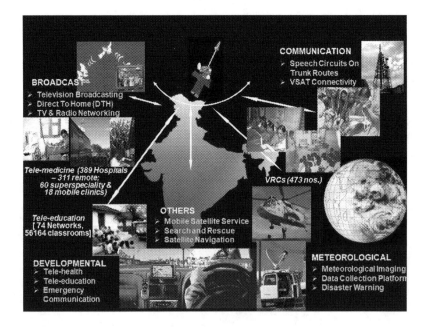

FIGURE 7.4   Electronic connectivity through space technology.

EARTH OBSERVATION SATELLITES

Earth observation satellites are intended for environmental monitoring, meteorology, cartography, and others for application in precision farming, forest cover, water beds, wasteland development, ice cover on the poles, weather monitoring, fishing,border monitoring, and also military spying. A typical application of Earth Observation Satellite is given in Figure 7.5.

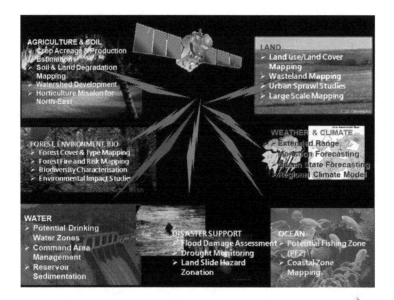

FIGURE 7.5   Earth observation satellite – Applications.

Starting with IRS-1A in 1988, ISRO has launched many operational remote sensing satellites, Currently, *thirteen* operational satellites are in Sun-synchronous orbit – RESOURCESAT-1, 2, 2A CARTOSAT-1, 2, 2A, 2B, RISAT-1 and 2, OCEANSAT-2, SARAL and SCATSAT-1, and *four* in Geostationary orbit- INSAT-3D, Kalpana & INSAT 3A, INSAT-3DR. Varieties of instruments have been flown on-board these satellites to provide necessary data in a diversified spatial, spectral, and temporal resolutions to cater to different user requirements in the country and for global usage. The data from these satellites are used for several applications covering agriculture, water resources, urban planning, rural development, mineral prospecting, environment, forestry, ocean resources, and disaster management. The data generated are processed and stored as databases and made available to millions of the people in India through organised state-level resource management system of ISRO. In addition, the remote sensing data is being shared by users from many countries, having multiple ground stations throughout the world. Remote-sensing satellite data can quantify and provide macro information in multiple areas.

NAVIGATION

Satellite navigation uses satellites to provide autonomous geo-spatial positioning. It allows small electronic receivers to determine their location (longitude, latitude, and altitude/elevation) to high precision (within a few centimetres to metres) using time signals transmitted along a line of sight by radio from satellites. Satnav systems operate independently of any telephonic or Internet reception, though these technologies can enhance the usefulness of the positioning information generated.

A satellite navigation system with global coverage may be termed a global navigation satellite system (GNSS). As of September 2020, the United States' Global Positioning System (GPS), Russia's Global Navigation Satellite System (GLONASS), China's BeiDou Navigation Satellite System (BDS), and the European Union's Galileo are fully operational GNSSs. Japan's Quasi-Zenith Satellite System (QZSS) is a (US) GPS satellite-based augmentation system to enhance the accuracy of GPS, with satellite navigation independent of GPS scheduled for 2023 (https://en.wikipedia.org/wiki/Satellite_navigation-cite_note-3). The Indian Regional Navigation Satellite System (IRNSS) plans to expand to a global version in the years to come. Global coverage for each system is generally achieved by a satellite constellation of 18–30 medium Earth orbit (MEO) satellites spread between several orbital planes. The actual systems vary, but use orbital inclinations of >50° and orbital periods of roughly 12 hours (at an altitude of about 20,000 km). (Figure 7.6).

Navigation Systems	Country	Operator	Type	Coverage
Global Positioning System (GPS)	United States	Air Force Space Command (AFSPC)	Military, civilian	Global
GLONASS	Russia	Russian Aerospace Defense Forces,VKO	Military	Global
BeiDou Navigation Satellite System (BDS)	China	China National Space Administration (CNSA)	Military, commercial	Global Operational (regionally)
Indian Regional Navigation Satellite System, IRNSS (Operational by 2016)	India	Indian Space Research Organisation (ISRO)	Military, civilian	Regional
Galileo (In development)	European Union	GSA, ESA	Civilian, commercial	Global
Quasi-Zenith Satellite System (QZSS) (In development)	Japan	Japan Aerospace eXploration Agency (JAXA)	Civilian	Regional

FIGURE 7.6   Navigation systems around the world.

Indian Regional Navigation Satellite Systems (IRNSS) – NavIC

The **NavIC** or **NAVigation with Indian Constellation** is an autonomous regional satellite navigation system developed by Indian Space Research Organisation (ISRO), as shown in Figure 7.7. The government of India approved the project in

FIGURE 7.7   IRNSS (ISRO) mission, parts and assembly.

Credit: ISRO publication.

May 2006 and consists of a constellation of seven navigational satellites. Three of the satellites are placed in the geostationary orbit (GEO) and the remaining four in the geosynchronous orbit (GSO) to have a larger signal footprint and lower number of satellites to map the region. It is intended to provide an all-weather absolute position accuracy of better than 7.6 m throughout India and within a region extending approximately 1500 km around it. An extended service area lies between the primary service area and a rectangle area enclosed by the 30th parallel south to the 50th parallel north and the 30th meridian east to the 130th meridian east, 1500–6000 km beyond borders.

The constellation was in orbit as of 2018. The "standard positioning service" will be open for civilian use, and a "restricted service" (an encrypted one) for authorised users (including military). There are plans to expand the NavIC system by increasing constellation size from 7 to 11.

SMALL SATELLITES

Smallsat is a satellite of low mass and size, usually under 500 kg. While all such satellites can be referred to as "small", different classifications are used to categorise them based on mass, as shown below. With large-scale miniaturisation in electronic systems and use of COTS, satellites can be built small with substantial cost reduction. The associated cost of launch is also less. Launch vehicle with less payload capability with quick turnaround time will find large-scale use to deliver multiple smaller satellites in orbits. Miniature satellites, especially in large numbers, may be more useful than fewer, larger ones for same purposes – for example, gathering of scientific data and radio relay. Large satellites if destroyed by ASAT or failed in orbit will lead to huge loss and non-availability in critical time. But a small satellite becomes nonfunctional in a constellation, the operational need will not be affected, and cost loss will be minimum. Miniaturised satellites allow for cheaper designs, standardisation, and ease of mass production, like the CubeSat. Multiple small satellites can be injected in different orbits with apogee motor restart capability and can be "piggyback" of the last stage of the launch vehicle.

Another major reason for developing small satellites is the opportunity to enable missions that a larger satellite could not accomplish, such as:

- Constellations for low data rate communications
- Using formations to gather data from multiple points
- In-orbit inspection of larger satellites
- University-related research
- Testing or qualifying new hardware before using it on a more expensive spacecraft

Space Environment Hazards

Space environment is not benign. Satellites are subjected to radiation hazards, solar flares, charged particles high-energy protons, and cosmic rays. These affects the performance of the satellite and its life. So, radiation hardening is done in critical parts and components, and subassemblies go through rigorous tests.

PRIVATE SPACE OPERATORS

Space-X

SpaceX – Space Exploration Technologies Corp. is an American aerospace manufacturer, space transportation services, and communications company headquartered in Hawthorne, California. SpaceX was founded in 2002 by Elon Musk with the goal of reducing space transportation costs to enable the colonisation of Mars. It is not only a disruptive launch provider for missions to the International Space Station (saving NASA millions).

It is also the largest commercial operator of satellites on the planet. SpaceX's achievements include the first privately funded liquid-propellant rocket to reach orbit (Falcon 1 in 2008), the first private company to successfully launch, orbit, and recover a spacecraft (Dragon in 2010), the first private company to send a spacecraft to the International Space Station (Dragon in 2012), the first vertical take-off and vertical propulsive landing for an orbital rocket (Falcon 9 in 2015), the first reuse of an orbital rocket (Falcon 9 in 2017), and the first private company to send astronauts to orbit and to the International Space Station (SpaceX Crew Dragon Demo-2 in 2020). SpaceX has flown and re-flown the Falcon 9 series of rockets over 100 times.

Starlink

It is a satellite Internet constellation operated by SpaceX providing satellite Internet access coverage to most of the Earth. The constellation has grown to over 1700 satellites through 2021, and will eventually consist of many thousands of mass-produced small satellites in low Earth orbit (LEO), which communicate with designated ground transceivers. While the technical possibility of satellite Internet service covers most of the global population, actual service can be delivered only in countries that have licensed SpaceX to provide service within any specific national jurisdiction. As of January 2022, the beta Internet service offering is available in 25 countries.

STAR SHIP

SpaceX is also developing Star ship, a privately funded, fully reusable, super heavy-lift launch system for interplanetary spaceflight. Star ship is intended to become the primary SpaceX orbital vehicle once operational, supplanting the existing Falcon 9, Falcon Heavy, and Dragon fleet. Star ship will be fully reusable and will have the highest payload capacity of any orbital rocket ever on its debut.

While the SpaceX operated **22%** of the world's operational satellites as of April 2020, it went on to launch an additional 175 satellites in the span of one month, from August to September 2020. Planet lab owns 246 satellites corresponding to 15%, Spire Global owns 89 satellites (5%), Iridium owns 78 satellites (5%), One Web owns 74 satellites (4%) and makes 51% of commercial satellites in operation.

BLUE ORIGIN

Blue Origin, LLC is an American privately funded aerospace manufacturer and suborbital spaceflight services company headquartered in Kent, Washington. Founded in 2000 by Jeff Bezos, the founder and executive chairman of Amazon, the company is led by CEO Bob Smith and aims to make access to space cheaper and more reliable through reusable launch vehicles.

Blue Origin is developing a variety of technologies, with a focus on rocket-powered vertical take-off and vertical landing (VTVL) vehicles for access to suborbital and orbital space (https://en.wikipedia.org/wiki/Blue_Origin-cite_note-11). Initially focused on suborbital spaceflight, the company has designed, built and flown multiple testbeds of its New Shepard vehicle at its facilities in Texas. Developmental test flights of the New Shepard, named after the first American in space Alan Shepard, began in April 2015, and flight testing is ongoing. Blue Origin rescheduled the original 2018 date for first passengers several times, and eventually successfully flew its first crewed mission on 20 July 2021.

There are many more, worldwide. Many came as start-ups. Some of them succeeded. With many space industries grooming government-owned space efforts are moving towards use of private ventures in spacefaring nations, including India.

OTHER TYPES OF SATELLITES

Biosatellites are satellites designed to carry living organisms, generally for scientific experimentation.

Killer satellites are satellites that are designed to destroy enemy warheads, satellites, and other space assets.

Crewed spacecraft (spaceships) are large satellites able to put humans into (and beyond) an orbit, and return them to Earth. (The Lunar Module of the U.S. Apollo program was an exception; in that it did not have the capability of returning human occupants to Earth.) Spacecraft including spaceplanes of reusable systems have major propulsion or landing facilities. They can be used as transport to and from the orbital stations.

Reconnaissance satellites are Earth observation or communications satellites deployed for military or intelligence applications. Very little is known about the full power of these satellites, as governments who operate them usually keep information pertaining to their reconnaissance satellites classified.

Recovery satellites are those provide a recovery of military, biological, space-production, and other payloads from orbit to Earth.

Space-based solar power satellites are proposed satellites that would collect energy from sunlight and transmit it for use on Earth or other places.

Space stations are artificial orbital structures that are designed for human beings to live on in outer space. A space station is distinguished from other crewed spacecraft by its lack of major propulsion or landing facilities. Space stations are designed for medium-term living in orbit, for periods of weeks, months, or even years.

Tether satellites are satellites that are connected to another satellite by a thin cable called a tether.

Weather satellites are primarily used to monitor Earth's weather and climate.

Landings on Moon and Rover on Mars (Figure 7.8).

FIGURE 7.8a   A replica of **Luna 9, the first spacecraft landed on the Moon. Landing date** 3 February 1966, 18:45:30 GMT (USSR). **Russia, China**.

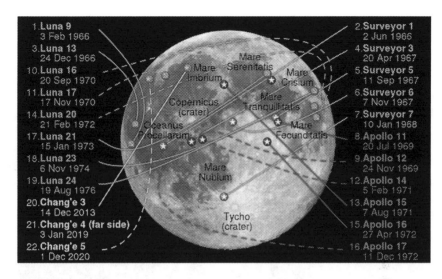

1.Luna 9		2.Surveyor 1
3 Feb 1966		2 Jun 1966
3.Luna 13		4.Surveyor 3
24 Dec 1966	Mare Serenitatis	20 Apr 1967
10.Luna 16	Mare Imbrium	5.Surveyor 5
20 Sep 1970	Mare Crisium	11 Sep 1967
11.Luna 17	Copernicus	6.Surveyor 6
17 Nov 1970	(crater) Mare Tranquillitatis	7 Nov 1967
14.Luna 20		7.Surveyor 7
21 Feb 1972	Oceanus Procellarum Mare Fecunditatis	10 Jan 1968
17.Luna 21		8.Apollo 11
15 Jan 1973		20 Jul 1969
18.Luna 23	Mare Nubium	9.Apollo 12
6 Nov 1974		24 Nov 1969
19.Luna 24		12.Apollo 14
19 Aug 1976		5 Feb 1971
20.Chang'e 3		13.Apollo 15
14 Dec 2013	Tycho	7 Aug 1971
21.Chang'e 4 (far side)	(crater)	15.Apollo 16
3 Jan 2019		27 Apr 1972
22.Chang'e 5		16.Apollo 17
1 Dec 2020		11 Dec 1972

FIGURE 7.8b   Landing near side of the Moon by USSR, China and USA.

FIGURE 7.8c   Apollo 11 Lunar Lander Eagle – 20 July 1969.

FIGURE 7.8d   Perseverance rover on Mars Feb 2021.

HUMAN SPACE MISSIONS

Yuri Gagarin was the first human to go to space and orbit around the Earth in April 1961. Rakesh Sharma was the first Indian to go to space in April 1984 in Soyuz (Figure 7.9a). **Sunita Williams in ISS** (Figure 7.9b). **Yang Liwei was the first Chinese astronaut to orbit the Earth in October 2003** (Figure 7.9c). **Chinese space station is already in orbit with space scientists** (Figure 7.9d).

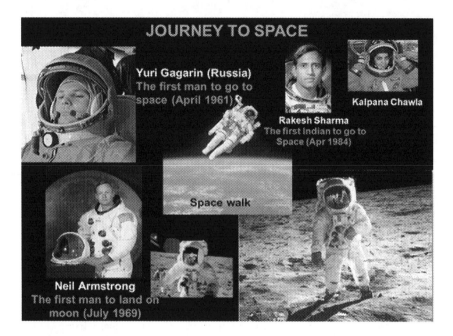

FIGURE 7.9a   Journey to space.

FIGURE 7.9b   Sunita Williams in International Space Station.

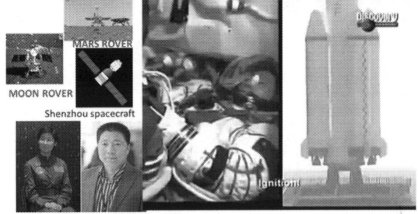

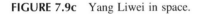

FIGURE 7.9c   Yang Liwei in space.

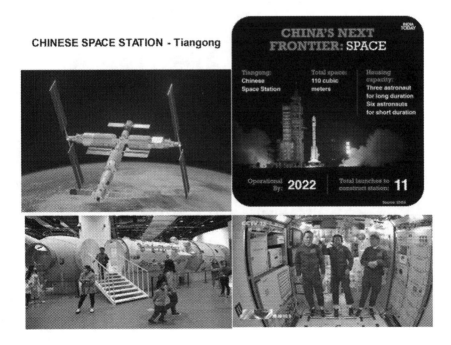

FIGURE7.9d   Prototype space station Chinese astronauts in space station.

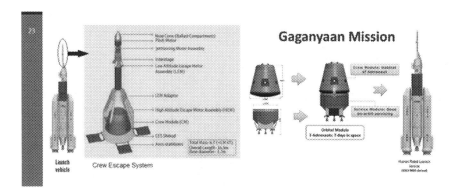

FIGURE 7.9e   India's human space program.

FIGURE 7.9f   Moon research station of China and Russia.

India is progressing all tasks for human space program to orbit Indian trained pilots in 2024 (Figure 7.9e). **Moon stations will become operational from 2024 first with Artemis by USA, followed by China + Russia** (Figure 7.9f).

8 Advances in Space Technologies

CAN WE TRAVEL FASTER THAN LIGHT?

Future space travel can be interplanetary between planets, interstellar between stars, and intergalactic between galaxies. The transport network for the interplanetary travel within our solar system could be a pathway gravitationally determined such that energy consumption is minimal. While travelling one could think of generating electricity to propel matter with little thrust operating for long time.

It could be with solar sails using the Sun's radiation or lasers, generating power from nuclear thermal or solar thermal engines for interstellar travel. Fusion rockets, anti-matter rockets, laser propulsion, and radiation propulsion could be used for deep space missions. Long trips of interstellar travel could extend the life span of human beings.

The special theory of relativity implies that only particles with zero rest mass may travel at the speed of light. **Tachyons,** particles whose speed exceeds that of light, have been hypothesised, but their existence would violate causality, and the consensus of physicists is that they do not exist. On the other hand, what some physicists refer to as "apparent" or "effective" faster than light depends on the hypothesis that unusually distorted regions of spacetime might permit matter to reach distant locations in less time than light could in normal or undistorted spacetime.

If we can make a bubble of spacetime travel faster than light, we could place a spaceship inside that bubble. Over 200 scientists from 13 countries have worked at OPERA (Oscillation Project with Emulsion-tRacking Apparatus) to find out whether Neutrinos can travel at a velocity greater than the speed of light in a vacuum.

DOI: 10.1201/9781003323396-8   **149**

The OPERA was an instrument used in a scientific experiment for detecting tau neutrinos from muon neutrino oscillations. The experiment is a collaboration between CERN in Geneva, Switzerland, and the Laboratori Nazionali del Gran Sasso (LNGS) in Gran Sasso, Italy, and uses the CERN Neutrinos to Gran Sasso (CNGS) neutrino beam.

The process started with protons from the Super Proton Synchrotron (SPS) at CERN being fired in pulses at a carbon target to produce pions and kaons. These particles decay to produce muons and neutrinos. After several experiments, from 2010, the OPERA collaboration updated their results. It was shown experimentally that the speed of neutrinos is consistent with the speed of light (https://en.wikipedia.org/wiki/OPERA_experiment-cite_note-op4–11). This was confirmed by a new, improved set of measurements in May 2013.

However, the universe is expanding faster than the speed of light. This gives a desire to know more about the influence of dark energy and dark matter to find out which particle can travel faster than light.

We will discuss various space technologies, including present electric propulsion used for satellites and the emerging space concepts, to increase the speed with high specific impulse. NASA and ESA have reported many new developments.

ELECTRIC PROPULSION

We discussed in the earlier chapters chemical propulsion rockets using solid, liquid, cryogenic propellants, a solid + liquid hybrid system, and air breathing propulsion. In this chapter, we will explore other types of propulsion systems, such as electric, thermal, plasma, laser, nuclear, and puled nuclear fusion propulsion.

THE NEED

Electric thrusters are the most efficient propulsion (operate at a higher specific impulse) unlike chemical rockets, which use stored energy in the form of propellants, leading to combustion and generating thrust. Electric thrusters use much less propellant, using stored electrical energy to generate thrust at a higher exhaust speed than chemical rockets. The propellant is ejected up to 20 times faster than from a classical chemical thruster, and therefore the overall system is many times more mass efficient. Due to limited electric power, the thrust is much less, but for a longer duration. So, it is suitable for spacecraft application, and very little mass is required to accelerate a spacecraft. The propellant used in these systems varies with the type of thruster and can generally be a rare gas (i.e., xenon or argon) or a liquid metal. However, electric propulsion is not a method suitable for launch vehicles, as they need larger thrust to leave Earth's gravity.

Electric propulsion is now a mature and widely used technology on spacecraft. Russian satellites have used electric propulsion for decades. As of 2019, over 500 spacecraft operated throughout the solar system used electric propulsion for station keeping, orbit raising, or primary propulsion. Particularly the communication satellites need orbit correction from the drift induced by solar radiation pressure and gravity gradients. Unlike chemical thrusters, electric thrusters (with milli newtons thrust) have the tremendous advantage of using less propellant with higher exhaust velocity, leading to a longer life for the satellite. Similarly, for transfer from GTO to GSO, electric thrusters give required apogee kick more efficiently. Nearly 30% weight is saved for the communication satellite, and life is extended in orbit, due to the use of electric thrusters.

In the early 1990s, NASA identified electric propulsion as a prime enabling technology for future deep space missions and began developing and testing various electric propulsion technologies. Intended to reduce fuel weight, decrease travel times to other planets, and permit larger scientific payloads, NASA started using electric propulsion technologies and also to continue exploration of Earth's neighbouring worlds. Electric propulsion technologies generate thrust via electrical energy that may be derived either from a solar source, such as solar photovoltaic arrays, which convert solar radiation to electrical power, or from a nuclear source, such as a space-based fission drive, which splits atomic nuclei to release large amounts of energy. This energy is used to accelerate an on-board propellant.

The basic electric propulsion system consists of three main components: some type of electric thruster that accelerates the ionised propellant, a suitable propellant that can be ionised and accelerated, and a source of electric power. The acceleration of electrically charged particles requires a large quantity of electric power.

TYPES OF ELECTRIC PROPULSION

Electric propulsion technologies typically considered for in-space use in Earth's orbit and beyond include:

- Plasma-based, xenon-fuelled Hall thrusters
- Solid Teflon-fuelled pulsed plasma thrusters
- Ammonia-fuelled arc jet thrusters
- Superheated water or nitrous oxide-fuelled resist jets
- Xenon-fuelled ion thrusters

The latter powers Deep Space 1, NASA's successful ion propulsion vehicle, which now cruises the solar system 220 million km from Earth, testing in-space hardware and electric propulsion capabilities.

Basically, there are three general types of electric rocket engines: electrothermal power, or the production of heat via electricity; electromagnetic power, or the production of magnetism via electricity; and electrostatic power, or the production of static electricity. These processes convert the accelerated propellant to spacecraft kinetic energy or thrust.

ELECTROTHERMAL ENGINE

In the basic **electrothermal rocket**, electric power is used to heat the propellant (such as ammonia) to a high temperature. The heated propellant is then expanded

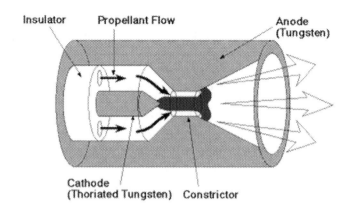

FIGURE 8.1   Electro thermal engine (DC arcjet).

through a nozzle to produce thrust. Propellant heating may be accomplished by flowing the propellant gas through an electric arc (this type of electric engine is called an arc jet engine) or by flowing the propellant gas over surfaces heated with electricity (Figure 8.1).

Although the arc jet engine can achieve exhaust velocities higher than those of chemical rockets, the dissociation of propellant gas molecules creates an upper limit on how much energy can be added to the propellant. In addition, other factors, such as erosion caused by the electric arc itself and material failure at high temperatures establish further limits on the arc jet engine. Because of these limitations, arc jet engines are more suitable for a role in orbital transfer vehicle propulsion and large spacecraft station keeping than as the electric propulsion system for deep space exploration missions.

RESITOJET

In a Resitojet propulsion system, propellant is heated by passing it over resistively heated surfaces and then through a converging-diverging nozzle (Figure 8.2).

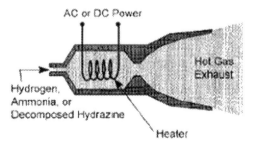

FIGURE 8.2   Resitojet engine.

ELECTROMAGNETIC/PLASMA ENGINE

The second major type of electric rocket engine is the electromagnetic engine or plasma rocket engine. In this type of engine, the propellant gas is ionised to form a plasma, which is then accelerated rearward by the action of electric and magnetic fields. The magneto plasma dynamic (MPD) engine can operate in either a steady state or a pulse mode. A high power (approximately 1 MW-electric) steady state MPD, using either argon or hydrogen as its propellant, is an attractive option for an electric propulsion orbital transfer vehicle (OTV).

MHD ENGINE

Pulsed Plasma Thrusters (PPTs) use a solid material (normally Teflon) for propellant, although very few use liquid or gaseous propellants. The first stage in PPT operation involves an arc of electricity passing through the fuel, causing ablation and sub-limation of the fuel. The heat generated by this arc causes the resultant gas to turn into plasma, thereby creating a charged gas cloud. Due to the force of the ablation, the plasma is propelled at low speed between two charged plates (an anode and cathode). Since the plasma is charged, the fuel effectively completes the circuit between the two plates, allowing a current to flow through the plasma. This flow of electrons generates a strong electromagnetic field, which then exerts a Lorentz force on the plasma, ac-celerating the plasma out of the PPT exhaust at high velocity (Figure 8.3).

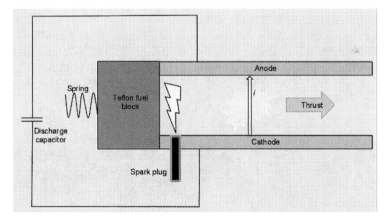

FIGURE 8.3   Schematic layout of a pulsed plasma thruster.

ELECTROSTATIC THRUSTERS

The third major type of electric rocket engine is the electrostatic rocket engine or ion rocket engine. As in the plasma rocket engine, Electric Propulsion for Deep-Space Missions. Propellant atoms (i.e., caesium, mercury, argon, or xenon) are ionised by removing an electron from each atom. In the electrostatic engine, however, the electrons are removed entirely from the ionisation region at the same rate as ions are accelerated rearward. The propellant ions are accelerated by an imposed electric field to a very high exhaust velocity. The electrons removed in the ioniser from the propellant atoms are also ejected from the spacecraft, usually by being injected into the ion exhaust beam. This helps neutralise the accumulated positive electric charge in the exhaust beam and maintains the ioniser in the electrostatic rocket at a high voltage potential.

An ion thruster or ion drive creates thrust by accelerating positive ions with electricity. The term refers strictly to gridded electrostatic ion thrusters and is often incorrectly loosely applied to all electric propulsion systems, including electromagnetic plasma thrusters.

ION THRUSTER

An ion thruster ionises a neutral gas by extracting some electrons out of atoms, creating a cloud of positive ions. These thrusters rely mainly on electrostatics as ions are accelerated by the Coulomb force along an electric field. Temporarily stored electrons are finally reinjected by a neutraliser in the cloud of ions after it has passed through the electrostatic grid, so the gas becomes neutral again and can freely disperse in space without any further electrical interaction with the thruster. Electromagnetic thrusters, on the contrary, use the Lorentz force to accelerate all species (free electrons as well as positive and negative ions) in the same direction, whatever their electric charge, and are specifically referred to as plasma propulsion engines, where the electric field is not in the direction of the acceleration.

Ion thrusters (Figure 8.4) in operational use have an input power need of 1–7 kW, exhaust velocity 20–50 km/s, thrust 25–250 mN, and efficiency 65%–80%, though experimental versions have achieved 100 kW, 5 N.

The Deep Space 1 spacecraft, powered by an ion thruster, changed velocity by 4.3 km/s while consuming less than 74 kg of xenon. The Dawn spacecraft broke the record, with a velocity change of 10 km/s.

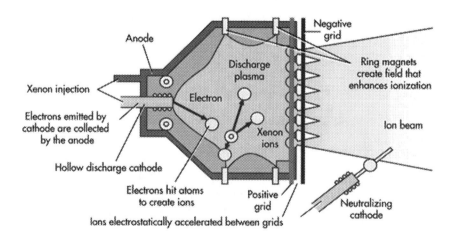

FIGURE 8.4 Schematic of a gridded ion thruster.

Applications include control of the orientation and position of orbiting satellites (some satellites have dozens of low-power ion thrusters) and use as a main propulsion engine for low-mass robotic space vehicles (such as Deep Space 1 and Dawn).

Ion thrust engines are practical only in the vacuum of space and cannot take vehicles through the atmosphere because ion engines do not work in the presence of ions outside the engine.

NASA Solar Technology Application Readiness (NSTAR) is a type of spacecraft ion thruster called electrostatic ion thruster. It is highly efficient low-thrust spacecraft propulsion running on electrical power generated by solar arrays. It uses high-voltage electrodes to accelerate ions with electrostatic forces.

HALL EFFECT THRUSTER

In spacecraft propulsion, a Hall-effect thruster (HET) is a type of ion thruster in which the propellant is accelerated by an electric field. Hall-effect thrusters (based on the discovery by Edwin Hall) are sometimes referred to as Hall thrusters or Hall-current thrusters. Hall-effect thrusters use a magnetic field to limit the electrons' axial motion and then use them to ionise propellant, efficiently accelerate the ions to produce thrust, and neutralise the ions in the plume. The Hall-effect thruster is classed as a moderate specific impulse (1600 seconds) space propulsion technology and has benefited from considerable theoretical and experimental research since the 1960s.

Hall thrusters operate on a variety of propellants, the most common being xenon and krypton. Other propellants of interest include argon, bismuth, iodine, magnesium, and zinc.

Hall thrusters are able to accelerate their exhaust to speeds between 10 and 80 km/s (1000–8000 seconds specific impulse), with most models operating between 15 and 30 km/s (1500–3000 seconds specific impulse). The thrust produced depends on the power level. Devices operating at 1.35 kW produce about 83 mN of thrust. High-power models have demonstrated up to 5.4 N in the laboratory. Power levels up to 100 kW have been demonstrated for xenon Hall thrusters.

As of 2009, Hall-effect thrusters ranged in input power levels from 1.35–10 kW and had exhaust velocities of 10–50 km per second, with thrust of 40–600 mN and efficiency in the range of 45%–60%. The applications of Hall-effect thrusters include control of the orientation and position of orbiting satellites and use as a main propulsion engine for medium-size robotic space vehicles.

Two types of Hall thrusters were developed in the Soviet Union:

1. Thrusters with wide acceleration zone, SPT (Stationary Plasma Thruster) at Design Bureau Fakel.
2. Thrusters with narrow acceleration zone, DAS (TAL, Thruster with Anode Layer), at the Central Research Institute for Machine Building (TsNIIMASH).

The SPT design was largely the work of A. I. Morozov. The first SPT to operate in space, SPT-50 aboard a Soviet Meteor spacecraft, was launched December 1971. They were mainly used for satellite stabilisation in north–south and in east–west directions. Since then, until the late 1990s, 118 SPT engines completed their mission, and some 50 continued to be operated. In 1982, SPT-70 and SPT-100 were introduced, their thrusts being 40 and 83 mN, respectively. In the post-Soviet Russia high-power (a few kilowatts) SPT-140, SPT-160, SPT-200, T-160, and low-power (less than 500 W) SPT-35 were introduced.

Soviet-built thrusters were introduced to the West in 1992 after a team of electric propulsion specialists from NASA's Jet Propulsion Laboratory, Glenn Research Center, and the Air Force Research Laboratory, under the support of the Ballistic Missile Defense Organization, visited Russian laboratories and experimentally evaluated the SPT-100 (i.e., a 100 mm diameter SPT thruster). Over 200 Hall thrusters have been flown on Soviet/Russian satellites in the past 30 years. No failures have ever occurred in orbit. Hall thrusters continue to be used on Russian spacecraft and have also flown on European and American spacecraft.

HET Principle of Operation

The essential working principle of the Hall thruster is that it uses an electrostatic potential to accelerate ions up to high speeds. In a Hall thruster, the attractive negative charge is provided by an electron plasma at the open end of the thruster instead of a grid. A radial magnetic field of about 100–300 G (0.01–0.03 T) is used to confine the electrons, where the combination of the radial magnetic field and axial electric field cause the electrons to drift in azimuth, thus forming the Hall current from which the device gets its name.

An electric potential between 150 and 800 V is applied between the anode and cathode.

The central spike forms one pole of an electromagnet and is surrounded by an annular space, and around that is the other pole of the electromagnet, with a radial magnetic field in between.

The propellant, such as xenon gas, is fed through the anode, which has numerous small holes in it to act as a gas distributor. As the neutral xenon atoms diffuse into the channel of the thruster, they are ionised by collisions with circulating high-energy electrons (typically 10–40 eV, or about 10% of the discharge voltage). Most of the xenon atoms are ionised to a net charge of +1, but a noticeable fraction (~20%) have +2 net charge.

The xenon ions are then accelerated by the electric field between the anode and the cathode. For discharge voltages of 300 V, the ions reach speeds of around 15 km/s (9.3 mps) for a specific impulse of 1500 seconds (15 kN·s/kg). Upon exiting, however, the ions pull an equal number of electrons with them, creating a plasma plume with no net charge.

The radial magnetic field is designed to be strong enough to substantially deflect the low-mass electrons, but not the high-mass ions, which have a much larger gyro radius and are hardly impeded. The majority of electrons are thus stuck orbiting in the region of high radial magnetic field near the thruster exit plane, trapped in $E \times B$ (axial electric field and radial magnetic field). This orbital rotation of the electrons is a circulating Hall current, and it is from this that the Hall thruster gets its name. Collisions with other particles and walls, as well as plasma instabilities, allow some of the electrons to be freed from the magnetic field, and they drift towards the anode.

About 20%–30% of the discharge current is an electron current, which does not produce thrust, thus limiting the energetic efficiency of the thruster; the other 70%–80% of the current is in the ions. Because most electrons are trapped in the Hall current, they have a long residence time inside the thruster and are able to ionise almost all of the xenon propellant, allowing mass use of 90%–99%. The mass use efficiency of the thruster is thus around 90%, while the discharge current efficiency is around 70%, for a combined thruster efficiency of around 63% (= 90% × 70%). Modern Hall thrusters have achieved efficiencies as high as 75% through advanced designs.

Compared to chemical rockets, the thrust is very small, on the order of 83 mN for a typical thruster operating at 300 V, 1.5 kW. As with all forms of electrically powered spacecraft propulsion, thrust is limited by available power, efficiency, and specific impulse. However, Hall thrusters operate at the high specific impulses that are typical for electric propulsion. One particular advantage of Hall thrusters, as compared to a gridded ion thruster, is that the generation and acceleration of the ions takes place in a quasi-neutral plasma, so there is no space charge saturated current limitation on the thrust density. This allows much smaller thrusters compared to gridded ion thrusters.

Another advantage is that these thrusters can use a wider variety of propellants supplied to the anode, even oxygen, although something easily ionised is needed at the cathode.

Propellants

Xenon

Xenon has been the typical choice of propellant for many electric propulsion systems, including Hall thrusters. Xenon propellant is used because of its high atomic weight and low ionisation potential. Xenon is relatively easy to store, and as a gas at spacecraft operating temperatures does not need to be vaporised before usage, unlike metallic propellants such as bismuth. Xenon's high atomic weight means that the ratio of energy expended for ionisation per mass unit is low, leading to a more efficient thruster.

Krypton

Krypton is another choice of propellant for Hall thrusters. Xenon has an ionisation potential of 12.1298 eV, while krypton has an ionisation potential of 13.996 eV. This means that thrusters utilising krypton need to expend a slightly higher energy per molecule to ionise, which reduces efficiency. Additionally, krypton is a lighter molecule, so the unit mass per ionisation energy is further reduced compared to xenon. However, xenon can be more than ten times as expensive as krypton per kilogram, making krypton a more economical choice for building out satellite constellations like that of SpaceX's Starlink, whose Hall thrusters are fuelled with krypton.

Cylindrical Hall Thrusters

Although conventional (annular) Hall thrusters are efficient in the kilowatt power regime, they become inefficient when scaled to small sizes. This is due to the difficulties associated with holding the performance scaling parameters constant while decreasing the channel size and increasing the applied magnetic field strength. This led to the design of the cylindrical Hall thruster. The cylindrical Hall thruster can be more readily scaled to smaller sizes due to its nonconventional discharge-chamber geometry and associated magnetic field profile. The cylindrical Hall thruster more readily lends itself to miniaturisation and low-power operation than a conventional (annular) Hall thruster. The primary reason for cylindrical Hall thrusters is that it is difficult to achieve a regular Hall thruster that operates over a broad envelope from ~1 kW down to ~100 W while maintaining an efficiency of 45%–55%.

External Discharge Hall Thruster

Sputtering erosion of discharge channel walls and pole pieces that protect the magnetic circuit causes failure of thruster operation. Therefore, annular and cylindrical Hall thrusters have a limited lifetime. Although magnetic shielding has been shown to dramatically reduce discharge channel wall erosion, pole piece erosion is still a concern. As an alternative, an unconventional Hall thruster design called external discharge Hall thruster or external discharge plasma thruster (XPT) has been introduced. External discharge Hall thruster does not possess any discharge channel walls or pole pieces. Plasma discharge is produced and sustained completely in open space outside the thruster structure, and thus erosion-free operation is achieved.

An illustration of the Gateway in orbit around the Moon is shown in Figure 8.5. The orbit of the Gateway will be maintained with Hall thrusters.

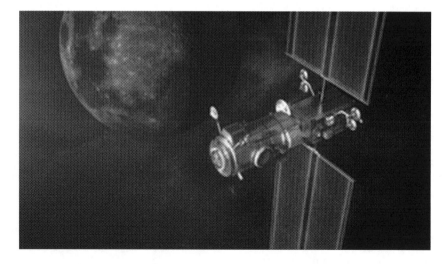

FIGURE 8.5   Illustration of the gateway.

Nanotechnology in Electric Propulsion Thrusters

Drastic miniaturisation of electronics and ingression of next-generation nanomaterials into space technology have provoked a renaissance in interplanetary flights and near-Earth space exploration using small satellites. The emerging satellite systems will use new designs that integrate nanomaterials to build advanced electric propulsion devices. (Ref. Recent progress and perspectives of space electric propulsion systems based on smart nanomaterials I. Levchenko et al, Nature Communications volume 9, Article number: 879 (28 February 2018)).

In the future, the most advanced electric thrusters may be able to impart a delta-v of 100 km/s, which is enough to take a spacecraft to the outer planets of the solar system at a faster speed. An electric rocket with an external power source (transmissible through laser on the photovoltaic panels) has a theoretical possibility for interstellar flight (Figure 8.6).

Other devices, such as magneto-plasma-dynamic thrusters, and pulsed induction thrusters, may offer future primary propulsion benefits for higher-power nuclear propulsion systems. Spacecraft powered by typical electric propulsion systems may

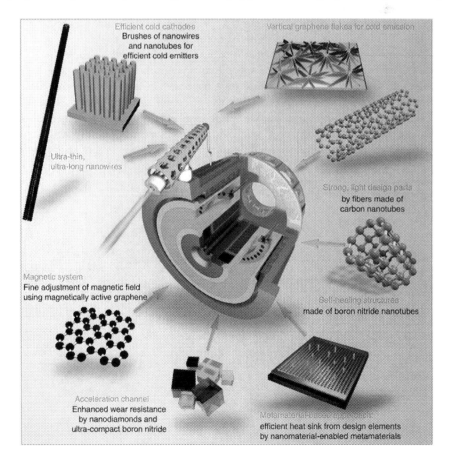

FIGURE 8.6 Future of electric thrusters using nanomaterials.

eject propellant at up to 20 times the speed of conventional chemical systems, delivering a much higher specific impulse, or the amount of thrust obtained for the weight of fuel burned. Electric-based systems also require far less propellant mass than traditional, chemical-propelled craft. In addition, deep-space missions no longer would be constrained by narrow launch windows dictated by planetary alignment. Traditionally, chemical-propelled spacecraft move from planet to planet as they travel, using "gravity-assist" manoeuvres in each world's orbit to increase their own velocity.

Magneto Plasma Rocket Propulsion (NASA)

The Variable Specific Impulse Magneto Plasma Rocket (VASIMR) is an electro-thermal thruster under development for possible use in spacecraft propulsion. It uses radio waves to ionise and heat an inert propellant, forming a plasma, then a magnetic field to confine and accelerate the expanding plasma, generating thrust. Schematic diagram is shown in Figure 8.7.

The VASIMR method for heating plasma was originally developed during nuclear fusion research. VASIMR is intended to bridge the gap between high thrust, low specific impulse chemical rockets and low thrust, high specific impulse electric propulsion, but has not yet demonstrated high thrust.

In these engines, a neutral, inert propellant is ionised and heated using radio waves. The resulting plasma is then accelerated with magnetic fields to generate thrust. Other related electrically powered spacecraft propulsion concepts are the electrodeless plasma thruster, the microwave arc jet rocket, and the pulsed inductive thruster. Every part of a VASIMR engine is magnetically shielded and does not directly contact plasma, increasing durability. Additionally, the lack of electrodes eliminates the electrode erosion that shortens the life of conventional ion thruster designs.

The propellant, a neutral gas such as argon or xenon, is injected into a hollow cylinder surfaced with electromagnets. On entering the engine, the gas is first heated to a "cold plasma" by a Helicon RF antenna/coupler that bombards the gas with

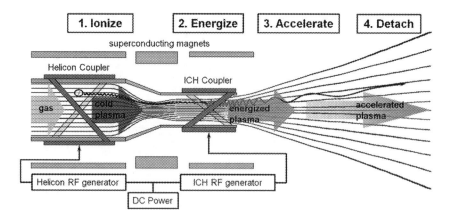

FIGURE 8.7   Schematic diagram of operation of VASIMAR engine.

electromagnetic energy, at a frequency of 10–50 MHz, stripping electrons off the propellant atoms and producing a plasma of ions and free electrons. By varying the amount of RF heating energy and plasma, VASIMR is claimed to be capable of generating either low-thrust, high–specific impulse exhaust or relatively high-thrust, low–specific impulse exhaust. The second phase of the engine is a strong solenoid-configuration electromagnet that channels the ionised plasma, acting as a convergent-divergent nozzle like the physical nozzle in conventional rocket engines.

A second coupler, known as the Ion Cyclotron Heating (ICH) section, emits electromagnetic waves in resonance with the orbits of ions and electrons as they travel through the engine. Resonance is achieved through a reduction of the magnetic field in this portion of the engine that slows the orbital motion of the plasma particles. This section further heats the plasma to greater than 1,000,000 K (1,000,000°C; 1,800,000°F) – about 173 times the temperature of the Sun's surface.

The path of ions and electrons through the engine approximates lines parallel to the engine walls; however, the particles actually orbit those lines while travelling linearly through the engine. The final, diverging, section of the engine contains an expanding magnetic field that ejects the ions and electrons from the engine at velocities as great as 50,000 m/s (180,000 km/h).

In contrast to the typical cyclotron resonance heating processes, VASIMR ions are immediately ejected from the magnetic nozzle before they achieve thermalised distribution. Based on novel theoretical work in 2004 by Alexey V. Arefiev and Boris N. Breizman of University of Texas at Austin, virtually all of the energy in the ion cyclotron wave is uniformly transferred to ionised plasma in a single-pass cyclotron absorption process. This allows for ions to leave the magnetic nozzle with a very narrow energy distribution, and for significantly simplified and compact magnet arrangement in the engine.

VASIMR does not use electrodes; instead, it magnetically shields plasma from most hardware parts, thus eliminating electrode erosion, a major source of wear in ion engines. Compared to traditional rocket engines with very complex plumbing, high performance valves, actuators and turbo pumps, VASIMR has almost no moving parts (apart from minor ones, like gas valves), maximising long-term durability.

High Power Electric Propulsion (HiPEP) is a variation of electrostatic ion thruster for use in nuclear electric propulsion applications. It was ground-tested in 2003 by NASA and was intended for use on the Jupiter Icy Moons Orbiter, which was cancelled in 2005.

The HiPEP thruster differs from earlier ion thrusters because the xenon ions are produced using a combination of microwave and magnetic fields. The ionisation is achieved through a process called Electron Cyclotron Resonance (ECR). In ECR, the small number of free electrons present in the neutral gas gyrate around the static magnetic field lines. The injected microwaves' frequency is set to match this gyro frequency and a resonance is established. Energy is transferred from the right-hand polarised portion of the microwave to the electrons. This energy is then transferred to the bulk gas/plasma via the rare – yet important – collisions between electrons and neutrals. During these collisions, electrons can be knocked free from the neutrals, forming ion-electron pairs. The process is a highly efficient means of creating a plasma in low density gases.

The thruster itself is in the 20–50 kW class, with a specific impulse of 6000–9000 seconds, and a propellant throughput capability exceeding 100 kg/kW. The pre-prototype HiPEP produced 670 mN of thrust at a power level of 39.3 kW using 7.0 mg/s of fuel giving a specific impulse of 9620 seconds. Downrated to 24.4 kW, the HiPEP used 5.6 mg/s of fuel giving a specific impulse of 8270 seconds and 460 milli N of thrust.

NUCLEAR PROPULSION ROCKETS FOR SPACE MISSIONS

A nuclear thermal rocket (NTR) is a type of thermal rocket where the heat from a nuclear reaction, often nuclear fission, replaces the chemical energy of the propellants in a chemical rocket. In an NTR, a working fluid, usually liquid hydrogen, is heated to a high temperature in a nuclear reactor and then expands through a rocket nozzle to create thrust. The external nuclear heat source theoretically allows a higher effective exhaust velocity and is expected to double or triple payload capacity compared to chemical propellants that store energy internally.

NTRs have been proposed as a spacecraft propulsion technology, with the earliest ground tests occurring in 1955. Although more than ten reactors of varying power output have been built and tested, as of 2021, no nuclear thermal rocket has been flown.

https://en.wikipedia.org/wiki/Nuclear_thermal_rocket-cite_note-unisci20190703-1

A NERVA Solid-core Design

Solid core nuclear reactors have been fuelled by compounds of uranium that exist in solid phase under the conditions encountered and undergo nuclear fission to release energy. Flight reactors must be lightweight and capable of tolerating extremely high temperatures, as the only coolant available is the working fluid/propellant. A nuclear solid core engine is the simplest design to construct and is the concept used on all tested NTRs.

A solid core reactor's performance is ultimately limited by the material properties, including melting point, of the materials used in the nuclear fuel and reactor pressure vessel. Nuclear reactions can create much higher temperatures than most materials can typically withstand, meaning that much of the potential of the reactor cannot be realised. Using hydrogen as a propellant, a solid core design would typically deliver specific impulses (I_{sp}) on the order of 850–1000 seconds, which is about twice that of liquid hydrogen-oxygen designs. Other propellants have also been proposed, such as ammonia, water, or LOX, but these propellants would provide reduced exhaust velocity and performance at a marginally reduced fuel cost. Yet, another mark in favour of hydrogen is that at low pressures it begins to dissociate at about 1500 K, and at high pressures around 3000 K. This lowers the mass of the exhaust species, increasing I_{sp}.

The concept for pulsed nuclear thermal rocket is for I_{sp} amplification. In this cell, hydrogen-propellant is heated by the continuous intense neutron pulses in the propellant channels. At the same time, the unwanted energy from the fission fragments is removed by a solitary cooling channel with lithium or other liquid metal.

The pulsed nuclear thermal rocket (not to be confused with nuclear pulse propulsion, which is a hypothetical method of spacecraft propulsion that uses nuclear explosions for thrust) is a type of solid nuclear thermal rocket for thrust and specific impulse (I_{sp}) amplification (Figure 8.8). In this concept, the conventional solid fission NTR can operate in a stationary as well as in a pulsed mode, much like a TRIGA reactor. Because the residence time of the propellant in the chamber is short, an important amplification in energy is attainable by pulsing the nuclear core, which can increase the thrust via increasing the propellant mass flow. However, the most interesting feature is the capability to obtain very high propellant temperatures (higher than the fuel) and then high amplification of exhaust velocity. This is because, in contrast with the conventional stationary solid NTR, propellant is heated by the intense neutron flux from the pulsation, which is directly transported from the fuel to the propellant as kinetic energy. By pulsing the core, it is possible to

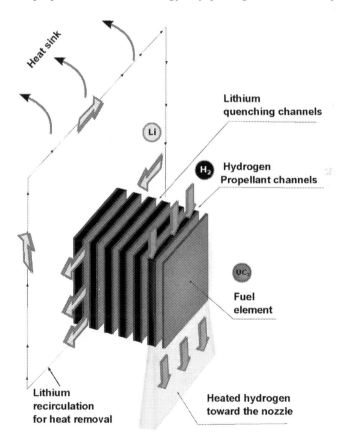

FIGURE 8.8   Pulsed nuclear thermal rocket.

obtain a propellant hotter than the fuel. However, and in clear contrast with classical nuclear thermal rockets (including liquid and gas nuclear rockets), the thermal energy from the decay of fission daughters is unwanted.

Very high instantaneous propellant temperatures are hypothetically attainable by pulsing the solid nuclear core, only limited by the rapid radiative cooling after pulsation.

Liquid Core

Liquid core nuclear engines are fuelled by compounds of fissionable elements in liquid phase. A liquid-core engine is proposed to operate at temperatures above the melting point of solid nuclear fuel and cladding, with the maximum operating temperature of the engine instead being determined by the reactor pressure vessel and neutron reflector material. The higher operating temperatures would be expected to deliver specific impulse performance on the order of 1300 to 1500 seconds.

A liquid-core reactor would be extremely difficult to build with current technology. One major issue is that the reaction time of the nuclear fuel is much longer than the heating time of the working fluid. If the nuclear fuel and working fluid are not physically separated, this means that the fuel must be trapped inside the engine while the working fluid is allowed to easily exit through the nozzle. One possible solution is to rotate the fuel/fluid mixture at very high speeds to force the higher density fuel to the outside, but this would expose the reactor pressure vessel to the maximum operating temperature while adding mass, complexity, and moving parts.

An alternative liquid-core design is the nuclear salt-water rocket. In this design, water is the working fluid and also serves as the neutron moderator. The nuclear fuel is not retained, which drastically simplifies the design. However, the rocket would discharge massive quantities of extremely radioactive waste and could only be safely operated well outside the atmosphere of Earth and perhaps even entirely outside the magnetosphere of Earth.

Gas Core – Closed Cycle and Open Cycle Figure 8.9

(a)   (b)

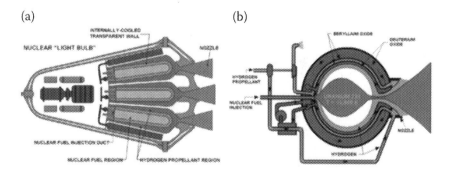

FIGURE 8.9   a) Nuclear gas core closed cycle rocket engine diagram, nuclear "light bulb" b) nuclear gas core open cycle rocket engineopen cycle rocket engine diagram.

The final fission classification is the gas-core engine. This is a modification to the liquid-core design, which uses rapid circulation of the fluid to create a toroidal pocket of gaseous uranium fuel in the middle of the reactor, surrounded by hydrogen. In this case, the fuel does not touch the reactor wall at all, so temperatures could reach several tens of thousands of degrees, which would allow specific impulses of 3000–5000 seconds. In this basic design, the "open cycle", the losses of nuclear fuel would be difficult to control, which has led to studies of the "closed cycle" or nuclear lightbulb engine, where the gaseous nuclear fuel is contained in a super-high-temperature quartz container, over which the hydrogen flows. The closed cycle engine has much more in common with the solid-core design, but this time is limited by the critical temperature of quartz instead of the fuel and cladding. Although less efficient than the open-cycle design, the closed-cycle design is expected to deliver a specific impulse of about 1500–2000 seconds (Figure 8.9).

BIMODAL NUCLEAR THERMAL ROCKETS

Bimodal Nuclear Thermal Rockets conduct nuclear fission reactions similar to those employed at nuclear power plants including submarines. The energy is used to heat the liquid hydrogen propellant. Advocates of nuclear-powered spacecraft point out that at the time of launch, there is almost no radiation released from the nuclear reactors. The nuclear-powered rockets are not used to lift off the Earth. Nuclear thermal rockets can provide great performance advantages compared to chemical propulsion systems. Nuclear power sources could also be used to provide the spacecraft with electrical power for operations and scientific instrumentation.

SOLAR ELECTRIC PROPULSION (SEP)

Energised by the electric power from on-board solar arrays, the electrically propelled system will use ten times less propellant than a comparable, conventional chemical propulsion system, such as those used to propel the space shuttles to orbit. Yet, that reduced fuel mass will deliver robust propulsion capable of boosting robotic and crewed missions well beyond low-Earth orbit: sending exploration spacecraft to distant destinations, ferrying cargo to and from points of interest, laying the groundwork for future missions, or resupplying those already underway.

With SEP technology, energy is fed into exceptionally fuel-efficient thrusters to provide gentle but nonstop thrust throughout the mission. The SEP project uses electrostatic Hall thrusters with advanced magnetic shielding – doing away with conventional chemical propellant delivered by a traditional rocket engine.

The thruster generates and traps electrons in a magnetic field, using them to ionise the onboard propellant – in this case, the inert gas xenon – into an exhaust plume of plasma that accelerates the spacecraft forward. Several Hall thrusters can be combined to increase power. A system able to accelerate xenon ions to more than 65,000 mph will provide enough force over time to move cargo and perform orbital transfers. SEP has been studied as a technology for a mission to Mars. In particular, the high specific impulse of the ion engines could lower overall mass and avoid having to use nuclear technology for power when coupled with solar panels (Figure 8.10).

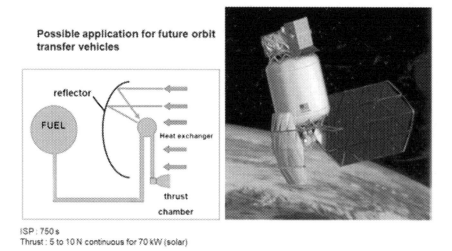

FIGURE 8.10   Solar thermal propulsion.

A fusion rocket is a theoretical design for a rocket driven by fusion propulsion, which could provide efficient and long-term acceleration in space without the need to carry a large fuel supply (Figure 8.11). The design relies on the development of fusion power technology beyond current capabilities, and the construction of rockets much larger and more complex than any current spacecraft. A smaller and

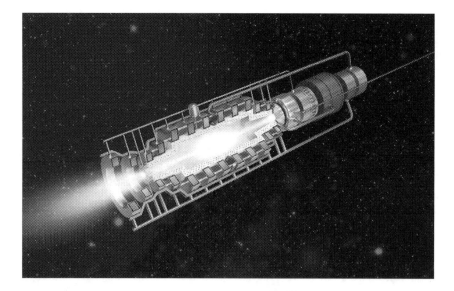

FIGURE 8.11   Fusion propulsion – Princeton satellite system.

lighter fusion reactor might be possible in the future when more sophisticated methods have been devised to control magnetic confinement and prevent plasma instabilities. Inertial fusion could provide a lighter and more compact alternative, as might a fusion engine based on a field-reversed configuration. Fusion nuclear pulse propulsion is one approach to using nuclear fusion energy to provide propulsion for rockets.

In the future, Helium-3 propulsion is a proposed method of spacecraft propulsion that uses the fusion of Helium-3 atoms as a power source. Helium-3, an isotope of helium with two protons and one neutron, could be fused with deuterium in a reactor. The resulting energy release could be used to expel propellant out the back of the spacecraft. Helium-3 is proposed as a power source for spacecraft mainly because of its abundance on the Moon.

LASER PROPULSION

In the last decades the principle of laser propulsion has been demonstrated by several laboratory experiments all over the world. The primary goal of these investigations was the development of alternative launch concepts for very small satellites (nanosats) from Earth's surface to low-Earth orbits (LEO). Unfortunately, it turned out, that the required laser power will not be available in the near future. The rapid progress in the field of solid-state laser technology triggered the development of powerful compact pulsed laser sources, even for applications in zero gravity environments and space. Equipped with this, new laser sources laser propulsion technology offers new concepts for position keeping and attitude control of satellites or satellite constellations in orbit. Beam control with active and adaptive optical systems enables long range or remote laser propulsion. A future step may be space missions with sampling probes to asteroids or small planets and the return on a tractor beam. Today, the main objective is the development and evaluation of precise laser thrusters in the range from 0.1 µN to 1 mN. Alternative micro propulsion concepts are absolutely essential for many missions with precise attitude and orbit control. There is a growing demand due to geodesic missions measuring Earth's gravity (successors of CHAMP, GRACE or GOCE), x-ray astronomy with telescopes built out of two satellites with highly constant distance, and astronomical missions with arrays of telescopes in a synthetic aperture architecture (Darwin). Due to its high precision and the simple (propellant) infrastructure laser propulsion is an ideal technology for micro thrusters. Precise adjustable thrust can be generated by laser-induced ablation of metals or composites with pulsed laser sources. Solar-pumped laser systems are an ideal technology for in-orbit propulsion.

SOLAR AND LASER POWERED INTERSTELLAR SAIL

Solar sail is a method of spacecraft propulsion using radiation pressure exerted by sunlight on large mirrors. The first spacecraft to make use of the technology was IKAROS, launched in 2010 (Figure 8.12).

FIGURE 8.12   IKAROS space-probe with solar sail in flight.

Courtesy: immeasurably.art (artist's depiction) showing a typical square sail configuration.

Solar sails use a phenomenon that has a proven, measured effect on astro-dynamics. Solar pressure affects all spacecraft, whether in interplanetary space or in orbit around a planet or small body. A typical spacecraft going to Mars, for example, will be displaced thousands of kilometers by solar pressure, so the effects must be accounted for in trajectory planning, which has been done since the time of the earliest interplanetary spacecraft of the 1960s. Solar pressure also affects the orientation of a spacecraft, a factor that must be included in spacecraft design.

The total force exerted on an 800 by 800 m solar sail, for example, is about 5 N at Earth's distance from the Sun, making it a low-thrust propulsion system, similar to spacecraft propelled by electric engines, but as it uses no propellant, that force is exerted almost constantly and the collective effect over time is great enough to be considered a potential manner of propelling spacecraft.

It could be even solar sails using laser, generating power from nuclear thermal or solar thermal engines and for interstellar travel.

Hypersonic Transportation

Many experiments have been carried to prove hypersonic vehicle for a reasonable duration by USA, Russia, Australia, and China. Operational missiles using kerosene have been inducted by Russia. While wave riders have shown good results for high-speed dive, long duration scramjet using hydrogen has not been achieved by any country for space application. In India, attempts have been made by ISRO and DRDO to prove scramjet (one flight each) for Mach 6 at an altitude of 20 km, as demonstration for a few seconds.

India proposed in the International Astronautical federation in 1988 at Bangalore, for the first time, a new concept of Hyperplane (Figure 8.13). Lead taken by Air Cmde Gopalaswamy, the team brought out that the payload efficiency can be improved to 15% compared to the 7% to 8%, proposed by the USA, Europe, and Japan. The idea is to develop a hyperplane vehicle that can take off from conventional airfields, collect air in the atmosphere on the way up, liquefy it, separate oxygen, and store it on board

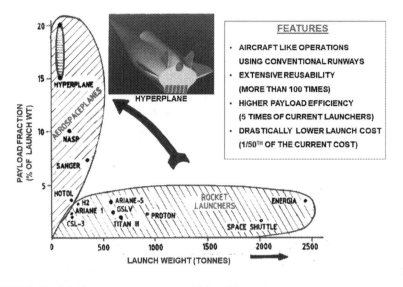

FIGURE 8.13 Multi-purpose aerospace vehicle – Hyperplane.

for subsequent flight beyond the atmosphere. It would take off horizontally like a conventional airplane from a conventional airstrip using turbo-ramjet engines that burn air and hydrogen. Once at a cruising altitude, the vehicle would use scramjet air breathing propulsion to accelerate from Mach 4 to Mach 8. During this cruising phase, an on-board heat exchange would collect the hot air from the engine and convert it into liquid oxygen. The liquid oxygen collected then would be used in the final flight phase when the rocket engine burns the collected liquid oxygen and the carried hydrogen to attain orbit. The vehicle would be designed to permit at least a 100 re-entries into the atmosphere. When operational, it is planned to be capable of delivering a payload weighing up to 1000 kg to low-Earth orbit. It would be the cheapest way to deliver material to space

This type of mission will be highly useful for multiple applications. In the case of Hyperplane, the aim was to achieve larger payload fraction to achieve lower cost per kg of satellite. The space shuttles of the USA with 2000 tonnes take-off weight could launch only 30 tonnes in the low-Earth orbit, giving a payload fraction of 1.5%. India's concept of Hyperplane aims to realise 15% of payload fraction (Figure 8.14). This will considerably reduce the launch cost per mission and will enable multiple missions such as transport, reconnaissance, payload delivery, satellite injection, etc.

On a typical mission, Hyperplane would take off with 100 tonnes weight using fan ramjet engine and then on scramjet mode for nearly 1000 seconds, during which time it collects the left-over air, cools it, and separates as liquid oxygen. This increases its weight to 166 tonnes; thereafter, it flies in rocket engine mode using the liquid oxygen and stored liquid hydrogen to deliver a payload of 16 tonnes. This concept of mass addition in flight is unique and has been conceived by Indian scientists. The mission profile is shown in Figure 8.14.

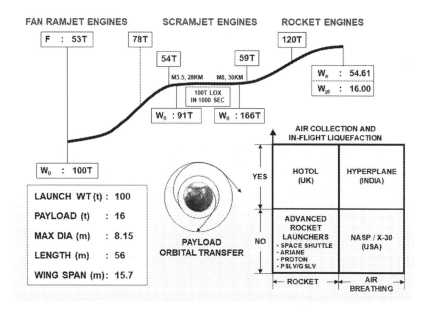

FIGURE 8.14   Hyperplane mission profile – Indian concept.

Low-Cost Access to Space

A cost analysis for access to space for different space missions of launching satellites using conventional expendable launch vehicles, orbital transportation by partial reusable vehicles like space shuttles, and large missions such as space colonisation using fully reusable Hyperplanes are shown in Figure 8.15. The price per

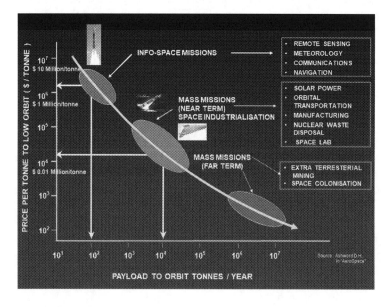

FIGURE 8.15   Cost of access to space.

tonne for the low-Earth orbit can be drastically reduced from $ 10 million to $1000 by going for the Hyperplane approach.

CLEAN ENERGY THROUGH SPACE

Clean energy is essential. Moreover, the availability of fossil fuels like oil and gas for power generation will be exhausted by 2075 and coal by 2100. Therefore, to meet the looming energy crisis and save Earth from pollution, the development and large-scale commercial utilisation of outer space has been suggested, with the construction of photovoltaic solar power satellites generating electric power for use on Earth. Solar energy is available for 99% of the time in an orbit above Earth, where 1.43 kW of solar energy illuminates any one square meter considerably greater than that received on Earth's surface (Figure 8.16).

Large solar power stations convert solar flux into microwave energy and beam it down to receiving stations at offshore locations on Earth. However, the construction of SPS in space would necessitate the use of Hyperplane, a heavy lift high-efficiency space cargo vehicle, using advanced aerospace technologies for revenue-earning mass missions in space. Studies have estimated that one SPS generating about 1000 MW would require 12 sq. km array of photovoltaic cells and would weigh 10,000 tonnes. Such SPS would take about three years for construction in space using a fleet of Hyperplanes to place construction materials in a low-Earth orbit.

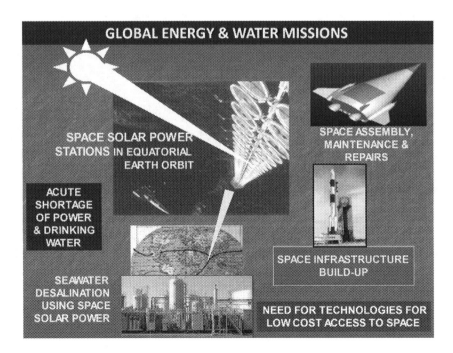

FIGURE 8.16   Global energy and water missions through solar power stations.

The Sun, as it is known, radiates about 10 trillion times the energy, which humans consume across the world today. If we can extract even a small portion of this energy from the sun, it would be sufficient to secure the energy demands of our future. Space-based solar power has many advantages over traditional terrestrial based solar plants. The level of Solar Irradiance is about 1.4 times in extra-terrestrial level than at the surface of the Earth. Collection time for the space-based solar power plant is 24 hours as compared to the 6–8 hours of surface-based solar power plants. This space-based solar power plant is not affected by the weather conditions, which may bring down the efficiency in the case of terrestrial power plants. Thus, space-based solar power plants would be far more effective in their efficiency and power generation than the land-based systems. There are three major focus areas in the space-based solar power plant. The first component is the space-based solar power plant; the second is the Earth-based collection system; and the third important aspect is the medium of transmission from space to Earth.

The aspect of safety and efficiency must be paramount in the way energy is transmitted from space back to Earth, either through microwave or any other technology like laser technology. Careful research of the impact and safety concerns would have to be conducted. One way to increase safety and improve efficiency could be the evolution of nano-packs, which are reusable, and can move like small batteries carrying charge back and forth from space solar station to ground reception. Another, approach could be to make the reception centres as pre-designated offshore sites to reduce the safety issues. Another important factor is the cost of the space-based power plant, which given the current launch technologies, would be very high and needs to be brought down. Among the largest cost components of installing a space based solar power plant will be the launching cost of the components into the orbit.

ARTIFICIAL INTELLIGENCE AND ROBOTIC SPACECRAFT DEVELOPMENT

The idea of using high-level automated systems for space missions has become a desirable goal to space agencies all around the world. Such systems are believed to yield benefits such as lower cost, less human oversight, and the ability to explore deeper in space, which is usually restricted by long communications with human controllers. Autonomy will be a key technology for the future exploration of the solar system, where robotic spacecraft will often be out of communication with their human controllers.

AUTONOMOUS SYSTEMS

Autonomy is defined by three requirements:

- The ability to make and carry out decisions on their own, based on information on what they sensed from the world and their current state.
- The ability to interpret the given goal as a list of actions to take.
- The ability to fail flexibly, meaning they are able to continuously change their actions based on what is happening within their system and their surroundings.

Currently, there are many projects trying to advance space exploration and space craft development using artificial intelligence (AI).

NASA's Autonomous Science Experiment

NASA began its autonomous science experiment (ASE) on Earth Observing-1 (EO-1), which is NASA's first satellite in the millennium program, Earth-observing series launched on 21 November 2000. The autonomy of these satellites is capable of on-board science analysis, re-planning, robust execution, and model-based diagnostic. Images obtained by the EO-1 are analysed on-board and down linked when a change or interesting event occurs. The ASE software has successfully provided over 10,000 science images. This experiment was the start of many that NASA devised for AI to impact the future of space exploration.

Artificial Intelligence Flight Adviser

NASA's goal with this project is to develop a system that can aid pilots by giving them real-time expert advice in situations that pilot training does not cover or just aid with a pilot's train of thought during flight. Based on the IBM Watson cognitive computing system, the AI Flight Adviser pulls data from a large database of relevant information like aircraft manuals, accident reports, and close-call reports to give advice to pilots. In the future, NASA wants to implement this technology to create fully autonomous systems, which can then be used for space exploration. In this case, cognitive systems will serve as the basis, and the autonomous system will completely decide on the course of action of the mission, even during unforeseen situations. However, in order for this to happen, there are still many supporting technologies required.

In the future, NASA hopes to use this technology not only in flights on Earth, but for future space exploration. Essentially, NASA plans to modify this AI flight Advisor for Longer range applications. In addition to what the technology is now, there will be additional cognitive computing systems that can decide on the right set of actions based upon unforeseen problems in space. However, in order to make it possible, there are still many supporting technologies that need to be enhanced.

Stereo Vision for Collision Avoidance

This project, NASA's goal is to implement stereo vision for collision avoidance in space systems to work with and support autonomous operations in a flight environment. This technology uses two cameras within its operating system that have the same view, but when put together offer a large range of data that gives a binocular image. Because of its duo-camera system, NASA's research indicates that this technology can detect hazards in rural and wilderness flight environments. Because of this project, NASA has made major contributions toward developing a completely autonomous UAV. Currently, Stereo Vision can construct a stereo vision system, process the vision data, make sure the system works properly, and lastly performs tests figuring out the range of impeding objects and terrain. In the future, NASA hopes this technology can also determine the path to avoid collision. The near-term goal for the technology is to be able to extract information from point clouds and place this information in a historic map data. Using this map, the technology could then be able to extrapolate obstacles and features in the stereo data that are not in the map data. This would aid with the future of space exploration where humans can't see moving, impeding objects that may damage the moving spacecraft.

BENEFITS OF AI

Autonomous technologies would be able to perform beyond predetermined actions. They would analyse all possible states and events happening around them and come up with a safe response. In addition, such technologies can reduce launch cost and ground involvement. Performance would increase as well. Autonomy would be able to quickly respond upon encountering an unforeseen event, especially in deep space exploration, where communication back to Earth would take too long. Space exploration could provide us with the knowledge of our universe as well as incidentally developing inventions and innovations. Travelling to Mars and farther could encourage the development of advances in medicine, health, longevity, transportation, and communications that could have applications on Earth.

ROBOTIC SPACECRAFT DEVELOPMENT

Changes in spacecraft development will have to account for an increased energy need for future systems. Spacecraft heading towards the centre of the solar system will include enhanced solar panel technology to make use of the abundant solar energy surrounding them. Future solar panel development is aimed at their working more efficiently while being lighter.

RADIOISOTOPE THERMOELECTRIC GENERATORS

Radioisotope Thermoelectric Generators (RTEG or RTG) are solid-state devices, which have no moving parts. They generate heat from the radioactive decay of elements, such as plutonium, and have a typical lifespan of more than 30 years. In the future, atomic sources of energy for spacecraft will hopefully be lighter and last longer than they do currently. They could be particularly useful for missions to the outer solar system, which receives substantially less sunlight, meaning that producing a substantial power output with solar panels would be impractical.

ESTABLISH LUNAR INDUSTRY

Space offers an enterprise for the future generation with the next Industrial Revolution. Availability of exotic resources and low-gravity manufacturing in the Moon and Mars have tremendous prospects for mankind. Mining in planets would need innovative methods for exploring, processing, and transporting large quantities of rare materials to Earth (Figure 8.17). The Moon could become a potential transportation hub for interplanetary travel and launch pads. The Moon's sky is

FIGURE 8.17   Space frontiers – *Lunar industry to mine He-3*.

clear to waves of all frequencies. With interplanetary communication systems located on the far side, the Moon would also shield these communication stations from the continuous radio emissions from Earth. Hence, the Moon has the potential to become a launch base for interplanetary travel and "Telecommunications Hub."

Man's quest for perennial sources of clean energy, such as solar and other renewable energies and thermonuclear fusion, would be filled through mining/exploration on the Moon. Large deposits of Helium-3 in the Moon and Mars provide a solution for future energy demand. Also, the dry ice deposits on the planets would be a source of fuel for rocket engines. Around 100 kg of Helium-3 would have a coal equivalent value of $140 million. Access to lunar Helium-3 at competitive cost potentially offers an environmentally benign means of helping meet an anticipated nine-fold or higher increase in energy demand by 2050. Samples collected in 1969 by Neil Armstrong during the first lunar landing showed that Helium-3 concentrations in lunar soil are at least 13 parts per billion (ppb) by weight. Levels may range from 20 to 30 ppb in undisturbed soils. But at a projected value of $1410 per g, 100 kg of Helium-3 would be worth about $141 million. The highest concentrations are in the lunar Maria; about half the He3 is deposited in 20% of the lunar surface covered by the Maria. It is believed that there are large deposits of He3 that have been deposited by solar wind in the lunar soil. Since the lunar soil has been stirred by collisions with meteorites, possible availability of He3 in the Moon would be down to depths of several metres. That 1 million metric tonnes of He3, reacted with deuterium, would generate about 20,000 terawatt-years (one trillion (10 to 12th power) watt-years) of thermal energy. Hence, the Helium-3 is the future.

Beyond 2025, one can imagine that space tourism will become popular, visits to the Moon and further to Mars using large boosters, reusable vehicles such as Hyperplane, and advanced propulsion offering cost-effective transportation. Space factories on the Moon and Mars will be built for mining rare resources such as Helium 3 for generating fusion energy. The Moon will also become the future launch station for exploring the space frontiers as the gravity of the Moon is 1/6th of the Earth, making highly efficient launch vehicles, with a higher payload fraction.

SPACE FRONTIERS

With the increasing demand for electric power, many nations will depend on a Solar Power Satellite (SPS), which could be built in space and launched in the geo-stationary orbit. The SPS will become the most cost-effective means of generating power. Establishment of a space colony at L4/L5 liberation points could be realised from the base station on the Moon. New findings of Earth-like habitable planets will further enhance the establishment of alternate habitat in the other planets (Figure 8.18).

FIGURE 8.18 Space frontiers for next 100 years.

"A day will come when humans settle in a space colony and in other planets beyond the solar system and ultimately conquer the Milky Way."

9 Environment: Protection of Earth and Sustainable Development

"We do not have the fantasy of competing with the economically advanced nations.But we are convinced that if we are to play a meaningful role nationally,we must be second to none in the application of advanced technologies to the real problems of man and society."

DR VIKRAM SARABHAI

As of now, Earth is the only planet to harbour life, suitable to the human race. Humankind must be kind enough to protect the Earth.

Dr. APJ Abdul Kalam's last speech at IIM Shillong, just a few minutes before his passing away, stressed the need for transforming our Planet Earth to make it "liveable". That means, we must live in a sustainable world.

CHALLENGES FOR THE SPACE COMMUNITY

While the world is actively involved in the futuristic research, development, and missions related to the exploration of the solar system and beyond, search for extra-terrestrial life, and further exploration of the Moon and Mars, we cannot overlook some of the challenges faced by humans and the space community:

The protection and sustainability of the environment of Earth for better living requires many areas of research like integrated study of the atmosphere, accurate forecasting and predictions of climate and weather, breakthroughs in earthquake forecasting, making green energy and finding solutions to water problems for the growing population for the world, and fast-track improvement of education and healthcare systems throughout the globe. According to a World Food Program (WFP) analysis, nearly 270 million people are expected to face acute food shortage this year, compared to 150 million before the pandemic. The analysis further concludes that the number of people on the brink of famine has increased to

DOI: 10.1201/9781003323396-9

41 million, compared to 34 million last year. They require an integrated global approach to the use of space technology solutions.

As Dr. Vikram Sarabhai stressed, the large space programs would call for certain "paradigm shifts" in developed nations to work together to bring the benefits of space to humanity as a whole, including three billion people in the underdeveloped nations. This is possible only if we have a strong cooperation among the nations with every rich nation contributing substantially to technology and resources. Strong international cooperation indeed can accelerate the application of space science and technology, leading to fast results for societal application and the security dimension in space. Such an accomplishment of a goal would enable taking up massive world care missions. For sustainable living, Earth must be protected from external threats (asteroids, space debris) and internal threats (global warming, nature's fury). Let us address the issues: a) threats from outer space and b) threats from human.

THREATS FROM OUTER SPACE

The planets of the Solar System were born in a violent storm of asteroid-like objects that began 4.6 billion years ago and lasted for roughly 500 million years to get stabilised in their respective orbits. Ironically, this process, which initially assisted in life's origin by seeding the Earth with precious organic compounds, now threatens it. The planets failed to consume all of the asteroids, and the planetary leftovers are still orbiting the Sun today. Most of them are confined to the "main belt" of asteroids, in between the orbits of Mars and Jupiter.

ESA's Infrared Space Observatory (ISO) showed that there might be as many as two million asteroids larger than one kilometre in this region of space (Figure 9.1). Gravitational nudges from the planets can push them out, causing them to fall towards the Sun, which means that they may cross Earth's orbit and potentially collide with our world. Large asteroids can make huge damage to the Earth when they hit with high velocity, like the one impacted 66 million years before eliminating nearly 75% of species.

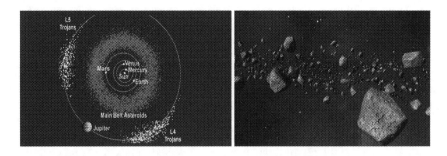

FIGURE 9.1   Asteroids in the belt.

Courtesy: The expanse wiki-fandom.

The Earth is in danger not only from asteroid strikes but also from meteors and comets. These comets usually live far away, beyond even Pluto, but can be jolted from their usual orbits by passing stars or gigantic gas clouds.

Beyond even this, "supernovae" could also pose a significant risk to Earth. A supernova is an exploding star that, for a few weeks, radiates more energy than all the other millions of stars in its galaxy put together. Such an occurrence takes place about once every century per galaxy. If the Earth were to be within 30 light years of a supernova, the violent radiation released would strip the Earth of its protective ozone layer, laying open the surface of our world to the full ravages of our Sun's ultraviolet light.

PROTECTION OF EARTH FROM RISKY ASTEROIDS

A collision 66 million years ago between the Earth and an asteroid approximately 10 km wide is thought to have produced the Chicxulub crater (located in what is now the Yucatán Peninsula in southeast Mexico), and triggered the Cretaceous–Paleogene extinction event that is understood by the scientific community to have caused the extinction of most dinosaurs, and also around 75% of the planet's animal species (according to new research by Southwest Research Institute (SwRI)).

While the chances of a major collision are low in the near term, it is a near-certainty that one will happen eventually unless defensive measures are taken.

Many times the asteroids fly past the Earth closely. All stray asteroids are being monitored. A global planetary defence system is getting evolved to provide early warning of dangerous asteroids larger than 40 m in size, about three weeks in advance, and able to deflect asteroids smaller than 1 km if known more than two years in advance. NASA and ESA have plans to launch several missions to divert asteroids, from hitting.

ASTEROIDS COLLISION AVOIDANCE STRATEGIES

Different collision avoidance techniques have trade-offs with respect to overall performance, cost, failure risks, operations, and technology readiness. There are various methods for changing the course of an asteroid. These can be differentiated by various types of attributes, such as the type of mitigation (deflection or fragmentation), energy source (kinetic, electromagnetic, gravitational, solar/thermal, or nuclear), and approach strategy (interception, rendezvous, or remote station).

Strategies broadly fall into two basic actions – i) **fragmentation and ii) deflection**. The asteroid can be split into fragments to scatter so that they miss the Earth or are small enough to burn up in the atmosphere. Deflection exploits the fact that orbits of both the Earth and the impactor meet at a point. Since the Earth is approximately 12,750 km in diameter and moves at approx. 30 km per second in its orbit, it travels a distance of one planetary diameter in about 425 seconds, or slightly over seven minutes. Delaying or advancing the impactor's arrival by times of this magnitude can, depending on the exact geometry of the impact, cause it to miss the Earth or deflect it to move to a different orbit.

The direct methods, such as nuclear explosives, or kinetic impactors, rapidly split the asteroid. Direct transfer energy to the object method is preferred because it is generally effective, less costly in time and money. Indirect methods, such as gravity tractors, attaching rockets or mass drivers, are much slower. They require travelling to the object, changing course up to 180° for space rendezvous, and then taking much more time to change the asteroid's path just enough so it will miss Earth.

A NASA analysis of deflection alternatives, conducted in 2007, stated:

> Nuclear standoff explosions are assessed to be 10–100 times more effective than the non-nuclear alternatives analyzed in this study. Other techniques involving the surface or subsurface use of nuclear explosives may be more efficient, but they run an increased risk of fracturing the target. They also carry higher development and operations risks.

In the same year, NASA released a study where the asteroid Apophis (with a diameter of around 300 m) was assumed to have a much lower rubble pile density (1500 kg/m^3) and therefore lower mass than it is now known to have, and in the study, it is assumed to be on an impact trajectory with Earth for the year 2029. Under these hypothetical conditions, the report determines that a "Cradle spacecraft" would be sufficient to deflect it from Earth impact. This conceptual spacecraft contains six B83 physics packages, each set for their maximum 1.2-megatonne yield, bundled together and lofted by an Ares V vehicle sometime in the 2020s, with each B83 being fused to detonate over the asteroid's surface at a height of 100 m ("1/3 of the objects diameter" as its stand-off), one after the other, with hour-long intervals between each detonation. The results of this study indicated that a single employment of this option "can deflect asteroids of 100–500 m diameter two years before impact, and larger ones with at least five years warning". These effectiveness figures are considered to be "conservative" by its authors, and only the thermal X-ray output of the B83 devices was considered, while neutron heating was neglected for ease of calculation purposes.

CLEARING THE SPACE DEBRIS

Space debris (also known as space junk, space pollution, space waste, space trash, or space garbage) is a term for defunct human-made objects in space – principally in Earth's orbit – which no longer serve a useful function. These include nonfunctional spacecraft and abandoned launch vehicle upper stages, mission-related debris, numerous fragmented debris near Earth, including those caused by ASAT tests (Figure 9.2). In addition to human-built objects left in orbit, other examples of space debris include fragments from their disintegration, erosion, and collisions, or even paint flecks, solidified liquids expelled from spacecraft, coolant released by nuclear-powered satellites, and unburnt particles from solid rocket motors.

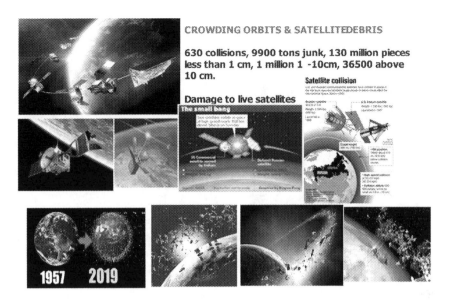

FIGURE 9.2   Satellite debris.

Courtesy: americanscientist.org, news.mit.com bussinessinaider.com etc.

The Haystack and HAX radars are located in Tyngsboro, Massachusetts. These radars collect 600 hours of orbital debris data each per year and are NASA's primary source of data on centimetre-sized orbital debris.

As the orbits of these objects often overlap the trajectories of spacecraft, debris is a potential collision risk. More than 800,000 debris, (25,000 are of larger size) are currently in orbit around Earth, in different orbits, each with the potential to cause damage to satellites in space. About 100 tonnes of this technological trash re-enters Earth's atmosphere in an uncontrolled way every year, the vast majority of which safely burns up – but some pieces make it through our atmosphere and reach the surface.

The passivation of spent upper stages by the release of residual fuels is aimed at reducing the risk of on-orbit explosions that could generate thousands of additional debris objects.

GEO ORBIT: Another process is self-removal. Geostationary satellites will be able to remove themselves to a "graveyard orbit" at the end of their lives. It has been demonstrated that the selected orbital areas do not sufficiently protect GEO lanes from debris, although a response has not yet been formulated. Rocket boosters and some satellites retain enough fuel to allow them to power themselves into a decaying orbit. In cases when a direct (and controlled) de-orbit would require too much fuel, a satellite can also be brought to an orbit where atmospheric drag would cause it to de-orbit after some years. Another proposed solution is to attach an electrodynamic tether to the spacecraft on launch. At the end of their lifetime, it is rolled out and slows down the spacecraft. It has also been proposed that booster

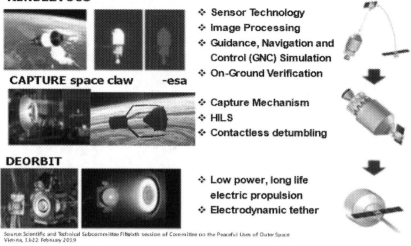

FIGURE 9.3 Active debris removal – Essential key technologies.

stages include a sail-like attachment to the same end. The vast majority of space debris, especially smaller debris, cannot be removed under its own power.

A variety of proposals have been made to directly remove such debris from orbit, by NASA, ESA, and by Russia and China. We need a well-coordinated integrated program to clean the space by the space faring nations. **The lead must be taken by ESA, bringing others under one umbrella like ITER or CERN.** These removals range from large spacecraft capture and hazard mitigation to "laser brooms" for removing small pieces of debris. Some of the possible methods for removal of debris are shown in Figure 9.3.

ESA Plan: By 2030, Europe, in a global effort with partners worldwide, will have a vibrant fleet of spacecraft in orbit around Earth, resilient to the hazards of space debris. Effort will include monitoring and safely managing the traffic in the "sky-ways." ESA will enable the safe operation of individual satellites and large constellations by developing and demonstrating an Automated Collision Avoidance System, free from causing damage.

Active Debris Removal/In-Orbit Servicing: ESA plans to develop an In Orbit Servicing Vehicle (IOSV) that will perform a variety of roles in orbit, including the ability to safely de-orbit satellites at the end of their lives. The new vehicle will also be able to refuel satellites, manoeuvre them, and ultimately demonstrate the technologies needed to extend the lifespan of missions from space.

ESA's Clean Sat initiative is to reduce the production of space debris – by developing technologies with consideration of the end of their lives in "design for demise" and promoting end-of-life passivation – emptying the tanks and discharging the batteries of satellites to prevent debris producing explosions. Different methods for the active removal of debris are shown in Figure 9.4.

ACTIVE REMOVAL OF DEBRIS

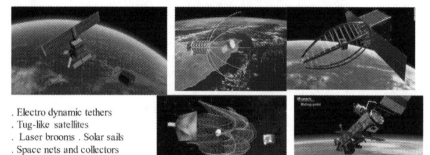

. Electro dynamic tethers
. Tug-like satellites
. Laser brooms . Solar sails
. Space nets and collectors

FIGURE 9.4   Active removal of debris.

NEW BUSINESS VENTURE

With the IOSV, ESA could spark off a valuable new business model for global industry that will go a long way toward mitigating new space debris and ensuring the long-term sustainability of spaceflight, taking a portion of large market opportunity in cleaning Earth space.

Hundreds of operational commercial and government satellites are in orbit, many of which will run out of fuel long before they sustain electronics or other systems failures. Their life could be extended by refuelling them, and thereby enormous cost can be saved. The goal is to find an optimum solution for refuelling, repairing, and servicing spacecraft in orbit. The advancements in robotics and artificial intelligence will give a way for performing such tasks in space. The humanoid robots could be deployed in space for undertaking any repair of the satellites and for extending the lifetime of the satellites. Space service can include activities such as asteroid redirection; satellite fleet management (traffic regulation); asteroid deflection/redirection; on-orbit assembly of platforms; on-orbit inspection & maintenance of observatories; space stations; propellant depots etc. This could be a business venture.

SPACE SYSTEMS AWARENESS AND COMPREHENSIVE SPACE SECURITY

In recent times, ground control systems for a number of NASA satellites were infiltrated and breached, while an ESA ground station experienced a so-called Denial of Service attack aimed at rendering it unable to relay data. Fortunately, no lasting damage was caused by these cyber-attacks, but these and other incidents highlight the importance of ensuring critical space systems do not fall victim to those with evil intent.

As on 1 January 2021, 3372 active satellites are in orbit around our planet, providing an enormous range of vital services, like navigation, telecommunication, survey of Earth's resources, and Internet access. Also, satellites are used for military applications. The USA has a Space Force to control and dominate in the space arena.

While big nations like the USA, Russia, and China are fighting to demonstrate supremacy over space, ESA is attempting to foster space safety and security efforts with the cooperation and support of 22 member states. Ensuring the safety and security of satellites is critical for civil society and developing nations. This includes not only protecting space infrastructure from space-based hazards, like space debris impacts, but also protecting from the cyber-security threats from Earth. Reliable and secure communication services are of great importance to understanding and reacting to the disasters, ensuring international coordination, and keeping emergency services, national and regional governments, and other civil organisations informed.

Considering space dominance by different countries, it is essential to create an International Space Force (ISF) made up of all space-faring nations to protect world space assets in a manner, which will enable peaceful use of space on a global cooperative basis. This can be under UN Security Council.

SPACE WEATHER

A proposal is being discussed in Europe to have a space weather monitoring system, to ensure the nations are resilient to threats from the Sun's emittance, solar radiation, flare, and meteors. This will include a dedicated space weather monitoring spacecraft, located at the fifth Lagrange, obtaining an excellent "side view" of the Sun; small satellites in orbit around Earth with space weather monitoring payloads; and piggy-back missions – monitoring instruments as "hosted payloads" with regular missions; and also, robust networks of space weather sensors on the ground.

MAN-CONTRIBUTED THREATS ON LAND

POPULATION INCREASE

Two factors are critical in life on Earth, first that the changing environment worst affects the poorest. Nearly, half of the population of the world lives within 150–200 m of the seacoast and factors like global warming are going to worst hit in these areas. Secondly, the economically weaker sections are most vulnerable to environment-borne diseases. Factors dominant for deterioration of environment are a) population growth in poor countries, b) increase of CO_2 concentration, and c) resulting global warming and increase in ocean temperature.

In 2025, the world population will reach 8 billion, and in 2050, it will cross nine billion (Figure 9.5). Africa, Asia, and South America nations with poor economy contribute this increase. More than three billion people will suffer for want of drinking water, clean air, and adequate healthcare. We have to find solution to overcome increasing dangers due to air and water pollution, increase in temperature

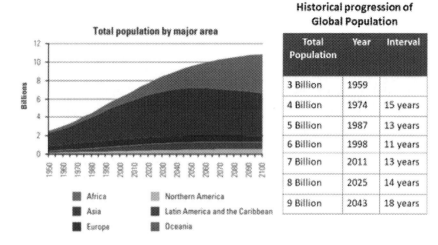

FIGURE 9.5 Population.

and also other threats to Earth from the fury of nature and falling of space debris and hitting of asteroids. Therefore, our priority must be to realise a Sustainable Development System using technology to ensure availability of adequate clean water and energy, increased agriculture products, and enhanced employment potential and empowerment of all the sectors, including rural economy.

GLOBAL WARMING

Global warming is a term used for the observed century-scale rise in the average temperature of the Earth's climate system and its related effects. The UN Intergovernmental Panel on Climate Change, a group of 1300 independent scientific experts from countries all over the world, in its Fifth Assessment Report concluded there is a more than 95% probability that human activities over the past 50 years have warmed our planet. Scientists believe that nearly all global warming is caused by increasing concentrations of greenhouse gases (GHGs) and other human-caused emissions.

Heat from the Earth is trapped in the atmosphere due to high levels of carbon dioxide (CO_2) and other heat-trapping gases that prohibit it from releasing the heat into space. This creates a phenomenon known today as the "greenhouse effect."

Long-lived gases that remain semi-permanently in the atmosphere and do not respond physically or chemically to changes in temperature are described as "forcing" climate change. Gases, such as water vapour, which respond physically or chemically to changes in temperature, are seen as "feedbacks." **It is estimated 51 billion tons of greenhouse gases, the world adds to the atmosphere every year.** Of course, Covid-19 has reduced this level by 5%–10%, at the cost of death of 20 million people and more than 1 billion people kept away from work. In spite of several attempts to reduce emission of greenhouse gases and carbon imprint, continuous human expansion resulted global warming when the atmosphere traps heat radiating from Earth toward space.

Gases that contribute to the greenhouse effect include:

Water vapour. The most abundant greenhouse gas, but importantly, it acts as feedback to the climate. Water vapour increases as the Earth's atmosphere warms, but so does the possibility of clouds and precipitation, making these some of the most important feedback mechanisms to the greenhouse effect.

Carbon dioxide (CO_2). A very important component of the atmosphere, carbon dioxide is released through natural processes such as respiration and volcano eruptions and through human activities such as deforestation, land use changes, and burning fossil fuels. Humans have increased atmospheric CO_2 concentration by 47% since the Industrial Revolution began. This is the most important long-lived "forcing" of climate change.

Methane. A hydrocarbon gas produced both through natural sources and human activities, including the decomposition of wastes in landfills, agriculture, and especially rice cultivation, as well as ruminant digestion and manure management associated with domestic livestock. On a molecule-for-molecule basis, methane is a far more active greenhouse gas than carbon dioxide, but also one which is much less abundant in the atmosphere.

Nitrous oxide. A powerful greenhouse gas produced by soil cultivation practises, especially the use of commercial and organic fertilisers, fossil fuel combustion, nitric acid production, and biomass burning.

Chlorofluorocarbons (CFCs). Synthetic compounds entirely of industrial origin used in a number of applications, but now largely regulated in production and release to the atmosphere by international agreement for their ability to contribute to destruction of the ozone layer. They are also greenhouse gases.

Over the last century, the burning of fossil fuels like coal and oil has increased the concentration of atmospheric carbon dioxide (CO_2). This happens because the coal or oil burning process combines carbon with oxygen in the air to make CO_2. To a lesser extent, the clearing of land for agriculture, industry, and other human activities has increased concentrations of greenhouse gases. Urban forests help to improve our air quality. Therefore, trees help by removing (sequestering) CO_2 from the atmosphere during photosynthesis to form carbohydrates that are used in plant structure/function and return oxygen back into the atmosphere as a by-product. Roughly half of the greenhouse effect is caused by CO_2. Therefore, trees act as carbon sinks, alleviating the greenhouse effect.

On an average, one acre of new forest can sequester about 2.5 tons of carbon annually. Young trees absorb CO_2 at a rate of 5 kg per tree each year. Trees reach their most productive stage of carbon storage at about ten years at which point they are estimated to absorb 20 kg of CO_2 per year. At that rate, they release enough oxygen back into the atmosphere to support two human beings. Planting 100 million trees could reduce an estimated 18 million tons of carbon per year. Planting trees remains one of the most cost-effective means of drawing excess CO_2 from the atmosphere. The World Watch Institute, in its *Reforesting the Earth* paper, estimated that the Earth needs at least 321 million acres of trees planted just to restore and maintain the productivity of soil and water resources, annually remove 780 million tons of carbon from the atmosphere and meet industrial and fuel wood needs in the third world. For every ton of new-wood growth, about 1.5 tons of CO_2 are removed from the air, and 1.07 tons of life-giving oxygen is produced.

CONSEQUENCES OF GLOBAL WARMING

Some effects are:

- On average, Earth will become warmer.
- Warmer conditions will probably lead to more evaporation and pre-cipitation overall, but individual regions will vary, some becoming wetter and others dryer.
- **A stronger greenhouse effect will warm the ocean and partially melt glaciers and ice sheets, increasing sea level. As of now, global sea level has risen by about 20 cm since reliable record keeping began in 1880. It is projected to rise another 2 m by 2100**. This is the result of added water from melting land ice and the expansion of seawater as it warms. In the next several decades, storm surges and high tides could combine with sea level rise and land subsidence to further increase flooding in many regions. **Many coastal cities, including New York, Mumbai, and Chennai, will have flooding, and parts will get submerged.**
- Outside of a greenhouse, higher atmospheric carbon dioxide (CO_2) levels can have both positive and negative effects on crop yields. Some labora-tory experiments suggest that elevated CO_2 levels can increase plant growth. However, other factors, such as changing temperatures, and water and nutrient constraints, may more than counteract any potential increase in yield. If optimal temperature ranges for some crops are exceeded, earlier possible gains in yield may be reduced or reversed altogether.
- Climate extremes, such as droughts, floods, and extreme temperatures, can lead to crop losses and threaten the livelihoods of agricultural producers and the food security of communities worldwide. Depending on the crop and ecosystem, weeds, pests, and fungi can also thrive under warmer temperatures, wetter climates, and increased CO_2 levels, and climate change will likely increase weeds and pests.
- Climate change can cause new patterns of pests and diseases to emerge, affecting plants, animals, and humans, and posing new risks for food se-curity, food safety, and human health.

OZONE DEPLETION

Due to the reasons stated above, ozone level has depleted. Already the ozone hole has appeared above Antarctica.

The global climate is projected to continue to change over this century and beyond. The magnitude of climate change beyond the next few decades depends

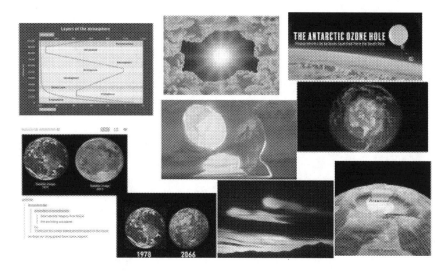

FIGURE 9.6   Ozone depletion in Antarctica.

primarily on the number of heat-trapping gases emitted globally, and how sensitive the Earth's climate is due to the ozone hole above Antarctica.

The **Arctic Ocean** is expected to become essentially ice free in summer before mid-century.

Ozone depletion in Antarctica is shown in Figure 9.6.

CO₂ CONCENTRATION

Accurately mapping the carbon balance of different regions of the world is necessary. This would include determining the carbon stock, forest and tree cover which offsets GHGs, agro-generated GHGs emissions, etc. This can accurately predict how much net GHG is being generated by each region and nation, which will be a significant improvement over the existing estimation methods being followed. Atmospheric carbon from fossil fuel burning is the main human-caused factor in the escalating global warming we are experiencing now. **The current level of carbon in our atmosphere is tracked using what is called the Keeling curve. The Keeling curve measures atmospheric carbon in parts per million (ppm).**

Each year, many measurements are taken at Mauna Loa, Hawaii, to determine the parts per million (ppm) of carbon in the atmosphere at that time. At the beginning of the Industrial Revolution (1) around 1880, before we began fossil fuel burning, our atmospheric carbon ppm level was at about 270.

Carbon dioxide (CO_2) is an important heat-trapping (greenhouse) gas, which is released through human activities such as deforestation and burning fossil fuels, as well as natural processes such as respiration and volcanic eruptions. Annual production of GHG has the following sources in percentage. Large-scale industry producing cement, steel, plastics etc 31%, Electricity 27%, Animal, plants growth 19%, Transportation – planes, ships, trucks etc. 16%, Household heating, refrigeration 7%. (ref. Bill Gates book)

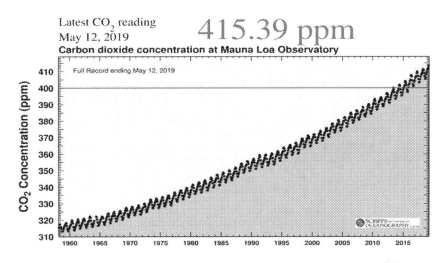

FIGURE 9.7 Variation of the concentration of mid-tropospheric CO_2 in parts per million.

Over the past 171 years, human activities have raised atmospheric concentrations of CO_2 by 48% above pre-industrial levels found in 1850. This is more than what happened naturally over a 20,000-year period (from the Last Glacial Maximum to 1850, from 185 ppm to 280 ppm).

The time series in Figure 9.7 shows global distribution and variation of the concentration of mid-tropospheric carbon dioxide in parts per million (ppm). The situation would be worsening with advancing time due to the annual increase of CO_2. The industrial activities have raised atmospheric carbon dioxide levels from 280 parts per million to 415 parts per million in the last 150 years. **In March 2021, the CO_2 level was 461 ppm.**

Catastrophic climate destabilisation is associated with a measurement of carbon 400–450 ppm. At the estimated current 1.2°C of temperature increase, we are already in the beginning stages of catastrophic climate destabilisation.

The *eventual* temperature *range* associated with catastrophic climate destabilisation will be an increase in the average global temperature of about 2.7°C. When global warming causes **storms, floods, seasonal disruption, wildfires, and droughts**, a nation is to spend $30–$100 billion per incident to repair. We are already in this phase of climate destabilisation, and many such incidents are happening.

TEMPERATURE INCREASE

Irreversible climate destabilisation is associated with a measurement beginning around carbon 425 ppm and going up to about carbon 550–600 ppm. The eventual temperature range associated with triggering irreversible climate destabilisation is an increase in average global temperature of 2.2°C–2.7°C to 4°C. Surface temperature will go up 3°C–4°C in 2100, with serious consequences. Rivers will become dry and scarcity for water, and depletion of the ozone layer will increase UV radiation.

Temperature Profile

Climate change will have the worst impact on poor farmers. We need to focus on new varieties of crops that tolerate the draughts and floods and stop adding greenhouse gases.

- Large-scale tree plantation is needed to absorb CO_2. – Renewable energy like wind, solar, and cultivation of biofuels can be encouraged.
- Throughout the world, the thermal power plants are functioning, and they generate tonnes and tonnes of fly ash. Can we work out a fly ash map and municipal waste map? Which will enable to convert fly ash into building material and fertiliser for effective utilisation of these resources?
- Thermal mapping of Earth due to global warming.
- Carbon dioxide is currently the most important greenhouse gas related to global warming. For the longest time, our scientists believed that once in the atmosphere, carbon dioxide remains there for about 100 years. New research shows that is not true: 75% of that carbon will not disappear for centuries to thousands of years. The other 25% stays forever. We are creating a serious global warming crisis that will last far longer than we ever thought possible.

Positive Actions

- Making carbon free electricity – solar, wind, geo-thermal, hydrogen, hydro, ocean tide and fusion energy, floating power stations with off-shore wind, sea-water conversion to drinking water, solar power satellite
- Electric vehicles only, power storage hydrogen, and nano-batteries
- Pollution absorbers in city roads
- Waste treatment plants to generate power
- Water treatment in rivers to eliminate water pollution, recycling of water at homes, offices
- Industrial waste, gas emittance to be treated
- Innovative methods of carbon capturing and clearing atmosphere
- Massive tree plantation in concrete jungles – one billion per year
- Promote new technology innovations, R&D Centres, incentives to achievers
- Government policies, regulations, standards, funding & infrastructure build-up to perfectly align with the above

UN Secretary-General has proposed six climate-positive actions for governments to take once they go about building back their economies and societies:

- Green transition: Investments must accelerate the decarbonisation of all aspects of our economy
- Green jobs and sustainable and inclusive growth
- Green economy: making societies and people more resilient through a transition that is fair to all and leaves no one behind
- Invest in sustainable solutions: fossil fuel subsidies must end, and polluters must pay for their pollution

- Confront all climate risks
- Cooperation – no country can succeed alone

To address the climate emergency, post-pandemic recovery plans need to trigger long-term systemic shifts that will change the trajectory of CO_2 levels in the atmosphere.

Governments around the world have spent considerable time and effort in recent years to develop plans to chart a safer and more sustainable future for their citizens. Taking these on board now as part of recovery planning can help the world build back better from the current crisis.

GEOSPATIAL TECHNOLOGIES FOR SUSTAINABLE DEVELOPMENT AND EMPOWERMENT

India has deployed successfully more than 25 remote sensing satellites (currently 13 in operation) in Sun synchronous orbits, for applications such as wet land and wasteland mapping, forest cover, urban planning, ocean resources analysis, crop information, border area monitoring, etc.

The data generated are processed and stored as database and made available to millions of the people in India through the organised state-level resource management system of ISRO. In addition, the remote-sensing data is being shared by users from many countries, having multiple ground stations throughout the world. Remote-sensing satellite data are able to quantify and provide macro information in multiple areas. Wasteland of 40 million hectares has been reported presently available, whereas the agriculture land used in the country is around 170 million hectares. One of the very important contributions of Earth observation is the evolution of a ground water map, which has been released recently. This single action has provided availability of water, and many citizens benefited. It is reported by users that water availability **is confirmed in 93%** cases while using Earth observation data. Not only for India, nearly three billion people who are below poverty need to be empowered through technology.

AGRICULTURE

About 38 out of every 100 workers in the world are into agriculture. In the least developed nations this ratio goes up to 68%. Our urgent challenge of this hour is how geospatial knowledge can enhance the potential of this workforce and uplift their lives. Can we think of:

a. Delivering better crop productivity by enhancing the spatial utilisation of the farmland, characterising the soil content, agro climatic variations, and changing weather conditions of the farmland? Can we manage the information from end-to-end for the crop cycle?
b. Mapping the water content of all water bodies, their silting status, and their potential to bring welfare to the farmlands. Can this be a verified, legally acceptable source of information which is regularly updated?

c. Better pest management and weed management by using GIS application to detect and map the spread of pests in the region and highlighting the vulnerable areas for urgent action.
d. One of the major challenges today is post-harvest management. This includes cold storages, silos, and markets for food and other farm produce. A geospatial community link can provide information over IVRS in local language for the farmers. This will greatly benefit the agriculture community. This geospatial community link can be undertaken as a social enterprise. There should not be delay for government actions to protect farmer's interest.
e. All this information can be made available in a farmer-friendly manner, in his language, and delivered over a mobile phone. India alone has about 200 million mobile phone users in rural areas, many of whom are farmers and fisherman. Thus, the communication network can empower the farmers and fishermen.

Precision Farming

This is a concept that implies observation, measurement, and response to inter- and intra-field variability in crops employing satellite data and information technology (IT).

The approach defines the crops and soil requirements for optimum productivity on the one hand and to preserve resources, ensure environmental sustainability, and protection on the other. This process into regular farming helps to solve the most vital problems in agriculture: resource wasting, high costs, and destructive environmental impact.

In recent years, the adoption of digital technologies in precision agriculture has been adjusting the ways that farmers treat crops and manage fields. One doesn't have to be an expert to see how the technology has changed the concept of farming making it more profitable, efficient, safer, and simple. Among other technologies, farmers have picked five they deem to be the best:

- GIS software and GPS agriculture
- Satellite imagery
- Drone and other aerial imagery
- Farming software and online data
- Merging datasets

As a result, modern farms get significant benefits from the ever-evolving digital agriculture. These benefits include reduced consumption of water, nutrients, and fertiliser; reduced negative impact on the surrounding ecosystem; reduced chemical runoff into local groundwater and rivers; better efficiency; reduced prices; and many more. Therefore, business becomes cost-effective, smart, and sustainable. Let's discuss some of these agricultural technologies.

Spatial images and tools for their interpretation enable farmers to distinguish the problem areas precisely, to decide what method to apply in the target zone, and to calculate the best time for that.

BASIC TECHNOLOGIES USED IN PRECISION FARMING

- **Variable rate technology (VRT)** – any technology or method allowing farmers to control the number of inputs applicable within defined farming areas. This technology uses specialised software, controllers, and differential global positioning system (DGPS). Basically, there are three approaches to VRT – manual, based on maps, or data from sensors.
- **GPS soil sampling** – this method is based on taking samples of soil to check nutrients, pH level, and other data to make profitable decisions in agriculture. Big data collected by sampling, is applied to calculate the variable rate for optimised seeding and fertilising.
- **Computer-based applications** – this refers to applications used to create precise farm plans, field maps, crop scouting, and yield maps and to define the exact amount of inputs to be applied to fields. Among the advantages of this method is the possibility to create an environmentally friendly farming plan, which in its turn helps to reduce the cost and increase yields. On the other side, these applications provide narrow value data that cannot be applied for big precision agriculture solutions due to the inability to integrate the obtained data into other supporting systems.
- **Remote sensing technology** – the method determines factors that can stress a crop at a specific time to estimate the amount of moisture in the soil. The dataset is obtained from drones and satellites. Compared to drone data, satellite imagery is more accessible and multi-purpose.

Precision agriculture enables remote field control and management employing sensors in fields proper as well as drones and satellites for surveillance from the sky. Satellite images seem to be the most lucrative option of remote sensing to begin with for on-line crop health monitoring, acquiring, processing, and analysing data. Block Development Office can store the complete information in one place, receive historical data and their comparative analysis, make reports, and share any needed information with all participants involved in the field management process (farmers, agronomists, agrarians in the field, insurance companies, traders, etc.), as in Figure 9.8.

FIGURE 9.8 Information availability on hand.

Information on Growth Stages

Relevant information on the plant growth cycle enables agronomists to competently choose the most appropriate time for field activities. This includes applications of fertilisers, insecticides, fungicides, distribution of irrigation, or drainage systems. Thus, one can tackle the problem at right on time.

Field Zoning Based on the Productivity Level

Since the fields on the farm differ in soil composition, nutrients required, the ability to retain water, and many other attributes, the best way is to apply the field zoning technique. It suggests a differential approach to determine the purpose of the land and the way to handle it.

This is done via almost real-time satellite images and comparative analysis of historical data in a certain retrospective. When you spot a certain similarity – it is likely to be regularity you can make work for you.

Internet of Things

The Internet of Things (IoT) and robotics replace humans in many spheres of life, and farming is no exception. As of now, multiple apps can calculate the amount of planting material or required nutrients per acre even more accurately than men would! We get crop conditions and weather forecasts via phone.

Autopilot machinery is smart enough to distinguish weeds from plants and ripe fruit from unripe ones.

Geographic Information Systems

Since fields are location-based, GIS software becomes an incredibly useful tool in terms of precision farming. While using GIS software, farmers are able to map current and future changes in precipitation, temperature, crop yields, plant health, and so on. It also enables the use of GPS-based applications in-line with smart machinery to optimise fertiliser and pesticide application; given that farmers don't have to treat the entire field, but only deal with certain areas, they are able to achieve conservation of money, effort, and time.

Another great benefit of GIS-based agriculture is the application of satellites and drones to collect valuable data on vegetation, soil conditions, weather, and terrain from a bird's-eye view. Such data significantly improves the accuracy of decision-making.

With crop monitoring, one can store the whole dataset in one place and get detailed and comprehensive analyses of weather conditions, plant development stages, the best amount and time for seeding or fertiliser applications, GIS field zoning, and much more. The smart software notifies about weather forecasts, crop conditions, and anomalies in their development enough in advance to prevent losses.

Equipped with relevant information and competent recommendations, we will be able to make the best of the farm, reducing the inputs of seeds and fertilisers and contributing to nature protection.

Digital Platforms employ satellite monitoring in order to speed up a farmer's decision-making so that he does not miss a crucial point of field treatment. Here are some of the features available in these platforms:

Crop monitoring allows the use of the Normalized Difference Vegetation Index (NDVI) for tracking crop health. Another important feature of crop monitoring is a Scouting app. It is both a mobile and desktop app that employs digital field maps. By analysing weather data in-line with the data on plant condition obtained from satellite imagery, farmers can precisely apply irrigation and prevent frost or heat damage.

Promising agricultural technologies are moving into the future by leaps and bounds. They offer substantial help for farmers in their endeavour for optimising inputs, simplifying farm management, and increasing productivity. Increased yields, as well as reduced maintenance costs, help boost profit margins.

FISHING

Indian National Centre for Ocean Information Services is providing information to the fishermen about the potential fishing zone. Information so provided is beneficial to the fishermen leading to a large catch. The centre is using the satellite data of the ocean temperature and colour (chlorophyll), twice a week

Based on this colour and temperature data, they can establish the potential fishing zone in different parts of the Indian coastline. This information is sent to each landing station in the coast. Details about the location of the potential fishing zone, its distance from the coast, the baring in which one has to go, and the depth of the zone are displayed for the fishermen (Figure 9.9). The data collected at the

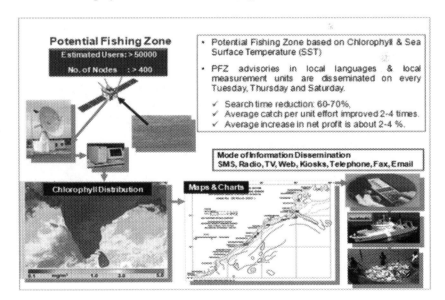

FIGURE 9.9   Potential fishing zone.

landing station may be communicated back to the fishermen inside the sea through SMS or other means of communication in the mode of radio waves or FM or through satellite communication. This information becomes vital tool for the fishermen in the region.

WATER AND LAND RESOURCES

The world's water resources are facing potential threats from various forms of water mismanagement either man made or due to natural calamity. Global climate change has changed dynamics of the environment resulting flood and drought in a changed dimensions and seasons and alters the geographic areas in land, river and coastal regions. Our water reserves in deep water table are being drawn down at alarming rates without being replenished. The quality of our water is being contaminated by pollutants, sediments, and sewage. Our river ways are becoming clogged with sediments due to erosion. These aspects need continuous observation for preservation, upkeep, and improvement. Certainly geospatial technology provides the means to monitor, measure, model, and manage these resources from the local to the global scale. It can be done nationally and globally.

Space-faring nations in the world have deployed many ocean satellites; India herself has a series of ocean satellites to explore all types of ocean wealth. These ocean satellites will facilitate the geo-governance through the creation of commonly accepted public policy by participating nations.

Interdisciplinary data collection in coastal upwelling regions, seafloor spreading centres, where tropical storms and hurricanes form, where oil spills occur, etc. will address many challenges, and the data will be collected from various platforms, instruments, at different study sites, at different scales and resolutions within these study sites. So, we are going to continue ways to organise, mine, and translate between data, which come from meta data. This will allow us to maintain and exchange data and information over large distances and long-time scales. Bhuvan provides a range of services enabling visualisation of various thematic data generated from various satellite resources.

BHUVAN – A UNIQUE GATEWAY TO INDIAN EARTH OBSERVATION DATA AND SERVICES

GIS and remote sensing are indeed an "enabling technologies" for marine science. We know the adage "location, location, location." But in the oceans, it is said "time is of the essence," as it is often, only by time that we can get location, especially on the deep seafloor or in the deeper parts of the water column that are out of reach of satellites, global positioning or otherwise. Accurate clocks and accurate timing of the travel of acoustic pulses are critical. In addition to that, Digital Multimedia

Broadcasting (DMB) Satellite (S-DMB) using a digital radio transmission technology can provide solution to many technology challenges, when coupled with remote sensing technologies and geo-spatial analysis on top of GIS.

GEO-SPATIAL PYRAMID

Visualisation of Geo-Spatial Pyramid structure linking data acquisition, information, value addition, knowledge, and wisdom and finally dissemination to the targeted users. They are called the User Community, which forms the bottom of the pyramid, and the user is the vital link for all economic activities.

Now, we need to refocus on how we are using technology of the 21st century to solve the problems, which are reminiscent of perhaps the 18th or 19th century. We need to re-think on how the geo-spatial technologies at our disposal can solve some of the problems of the three billion people in the rural population of the world and help them unleash their potential, thereby leading to a better human life, without damaging the environment around us.

Another challenge we face today is to take urban quality amenities to the three billion people in the rural population of the world. This is an urgent challenge, which will bridge the divide between the rich and poor, and urban and rural. The poorest of the world are paying the highest per unit cost for basic amenities of clean water, nutritious food, and healthcare. How can we overcome this ironic reality of the 21st century? Can the Geospatial Community champion the following missions?

a. Helping identify the state of potable water availability in the regions. This can include the parameters of both over- and underground water supply, pollution status, water-borne disease patterns, and usage data. This can be the basis for the **Potable Water GRID.**
b. Helping identify "hot spots" for local energy generation capacity. This can include energy from waste, energy from biofuels, which can be grown in wastelands, small-scale hydro plants, etc., which can empower the local communities. This can be the basis for the **Local Energy GRID.**

CLEAN ENERGY GENERATION

Another important aspect is energy security with freedom from fossil fuel. Around 86% of the total energy produced comes from fossil fuel; around 14% comes from renewable energy and the nuclear sector. In this situation, it is essential to find innovative methods to reduce the consumption of the electric power from fossil fuel and increase the deployment of renewable energy systems.

Solar, wind, hydrogen, and ocean based-current waves, and thermal gradients could form the future renewable power sources (Figure 9.10).

FIGURE 9.10   Clean energy generations – Sources.

India has 900 million mobile users, and 250,000 cell phone towers, which consumes nearly 2 billion litres of diesel for power. If we convert these installations into solar-powered systems, we save about 1.7 billion dollars and offset 5 million tonnes of CO_2 emission and gain carbon credit. Next, if we transform all our 600,000 villages where 700 million people live into solar-powered homes and streetlights, we may offset around 60% of fossil fuel usage in that sector.

10 Exoplanets, Earth 2.0

"If you do not have a mission, no problem will occur, but if you do have a mission or task definitely problems of varying magnitudes will crop up. But problems should not become the master of the individuals, individuals should become the master of the problem, defeat it and succeed."

~ Satish Dhawan

Various events in the past millions of years established that Earth is constantly being attacked from space and nature. A large asteroid of 10 km hit the Earth at the Yucatan Peninsula of Mexico 66 million years before sending into the sky debris that showered back all over Earth and eliminated dinosaurs and many other species. About 75,000 years before humanity, Asia and Africa almost got eliminated when a titanic explosion got triggered in Indonesia, spreading a thick blanket of poisonous ash, smoke, and debris. In another 1 million years, Earth will also become like Mars, a barren place or frozen with thick ice, not liveable. Nature will eventually turn on humanity and eliminate the human race from Earth, as it did to all those extinct 99.99% lifeforms. It is inescapable, as this is the norm of nature. Another long-term event to take place is the Sun exploding to become a red star, eliminating everything. Can we become a master of the problem? Can we settle in a planet outside the solar system?

If our long-term survival is at stake, we have a basic responsibility to our species to venture to other worlds.

—CARL SAGAN

Nearly 5000 exoplanets in the living zone have been located through telescopes. Some of them are described here in this chapter.

EXOPLANETS IN LIVEABLE ZONES

KEPLER-186f

Astronomers have discovered a number of Earth-size planets orbiting nearest stars with respect to the Sun, in the "habitable zone". One such planet first discovered

DOI: 10.1201/9781003323396-10

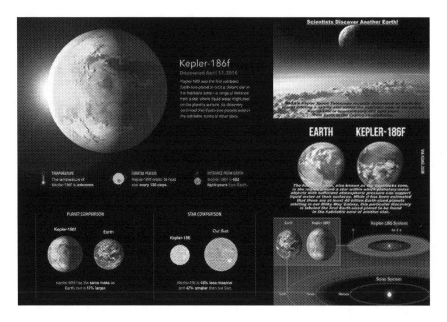

FIGURE 10.1   Kepler-186f.

was Kepler-186f (Figure 10.1), where liquid water might pool on the surface of the orbiting planet. The discovery of Kepler-186f confirms that planets the size of Earth exist in the habitable zone of stars other than our Sun.

Description

Kepler-186f is an exoplanet orbiting the red dwarf Kepler-186, about 582 light years from the Earth. It is the first planet with a radius similar to Earth's to be discovered in the habitable zone of another star.

> **Distance to Earth:** 557.7 light years
> **Radius:** 7454.1 km
> **Orbital period:** 130 days
> **Discovered:** 17 April 2014
> **Discoverer:** Elisa Quintana
> **Temperature:** 188 K (−85 °C)

KEPLER-62f

Using observations with NASA's Kepler Space Telescope, astronomers have discovered two new planetary systems that include three super-Earths.

Kepler-62f is a super-Earth-size planet in the habitable zone of a star smaller and cooler than the Sun (NASA Ames / JPL-Caltech), as in Figure 10.2a

The Kepler-62 planetary system, located about 1,200 light years from Earth in the constellation Lyra, hosts five exoplanets – 62b, 62c, 62d, 62e and 62f.

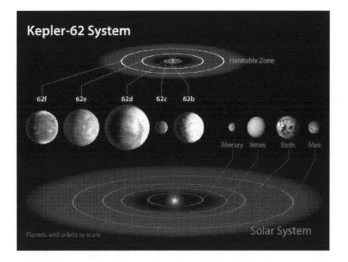

FIGURE 10.2a   Kepler-62 and the solar system (NASA Ames/JPL-Caltech).

Four of them are so-called super-Earths, larger than our own planet, but smaller than even the smallest giant planet in our solar system.

KEPLER-62e

Kepler-62e is a super-Earth-size planet in the habitable zone of a star smaller and cooler than the Sun (NASA Ames / JPL-Caltech) (Figure 10.2b).

FIGURE 10.2b   Kepler-62e.

These exoplanets have radii of 1.3, 1.4, 1.6, and 1.9 times that of Earth.

According to a paper published online in the journal *Science*, the two super-Earths in Kepler-62 (62e and 62f) orbit their star at distances where they receive about 41% and 120%, respectively, of the warmth from their star that the Earth receives from the Sun.

The planets are thus in the star's habitable zone; they have the right temperatures to maintain liquid water on their surfaces and are theoretically hospitable to life. Theoretical modelling suggests that both could be solid, either rocky or rocky with frozen water.

"This appears to be the best example our team has found yet of Earth-like planets in the habitable zone of a Sun-like star," said Dr Alan Boss of Carnegie Institution of Washington, co-author of the *Science* paper reporting the discovery of Kepler-62 exoplanets.

KEPLER-69c

The Kepler-69 system, located about 2700 light years away in the constellation Cygnus, hosts two planets – 69b and 69c, as in Figure 10.3.

Kepler-69c is 70% larger than the size of Earth and orbits in the habitable zone of Kepler-69. Astronomers are uncertain about the composition of this exoplanet, but its orbit of 242 days around a Sun-like star resembles that of our neighbouring planet Venus.

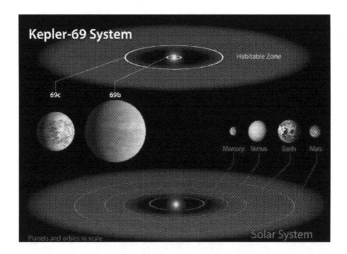

FIGURE 10.3 Kepler-69 and the solar system (NASA Ames/JPL-Caltech) Kepler 69 b is also called KOI 172.02 planet.

KEPLER-22b

Kepler-22b, also known by its Kepler object of interest designation KOI-087.01, is an extrasolar planet orbiting within the habitable zone of the Sun-like star

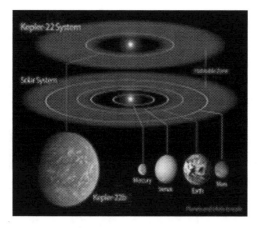

FIGURE 10.4   Kepler-22b.

Kepler-22. It is located about 638 light years from Earth in the constellation of
Cygnus (Figure 10.4).

Distance to Earth: 587.1 light years
Radius: 15,290 km
Orbital period: 290 days
Discovered: 5 December 2011
Temperature: 295 K (22°C; 71°F)
Detection method: Transit

TRAPPIST-1

Trappist-1 is an ultra-cool red dwarf star with a radius slightly larger than the planet
Jupiter, while having 84 times Jupiter's mass. It is located 39.6 light years from the
Sun in the constellation Aquarius.

Distance to Earth: 39.46 light years
Surface temperature: 2550 K
Radius: 84,180 km
Mass: 1.77×10^{29} kg (0.089 M☉)
Coordinates: RA 23h 6m 29s I Dec -5° 2′ 29″
Luminosity (visual, L_V): 0.00000373 L

Astronomers using NASA's Spitzer Space Telescope and ground-based telescopes discovered that the system has seven planets. Three of these planets are in
the theoretical "habitable zone," the area around a star where rocky planets are most

likely to hold liquid water. This landmark finding was announced on 22 February 2017; visual NASA's Hubble Space Telescope was used to find that TRAPPIST-1b and c were unlikely to have hydrogen-dominated atmospheres like those we see in gas giants.

The age of a star is important for understanding whether planets around it could host life. Scientists wrote in an August 2017 study that TRAPPIST-1 is between 5.4 and 9.8 billion years old. This is up to twice as old as our own solar system, which formed some 4.5 billion years ago. The Trappist-1 planets are shown in Figures 10.5 and 10.6.

TRAPPIST-1 Earth-size planets have been confirmed to be rocky. Assisted by several ground-based telescopes, including the European Southern Observatory's Very Large Telescope, Spitzer confirmed the existence of seven planets. Using Spitzer data, the team precisely measured the sizes of the seven planets and developed first estimates of the masses of six of them, allowing their density to be estimated. Based on their densities, all the TRAPPIST-1 planets are likely to be rocky. Further observations will not only help determine whether they are rich in water, but also possibly reveal whether any could have liquid water on their surfaces. The mass of the seventh and farthest exoplanet has not yet been estimated – scientists believe it could be an icy, "snowball-like" world, but further observations are needed.

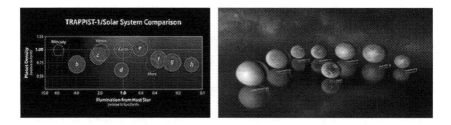

FIGURE 10.5 Relative sizes, densities, and illumination of the TRAPPIST-1 system compared to the inner planets of the solar system.

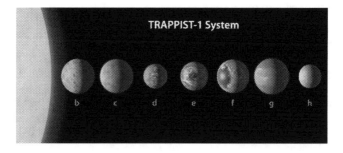

FIGURE 10.6 TRAPPIS-1 system.

New discovery in 2020

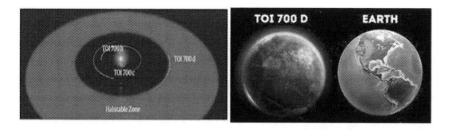

TOI 700 d is a near-Earth-sized exoplanet, likely rocky, orbiting within the habitable zone of the red dwarf TOI 700, the outermost planet within the system. It is located roughly 101.4 light years (31.1 pc) away from Earth in the constellation of Dorado. The exoplanet is the first Earth-sized exoplanet in the habitable zone discovered by the Transiting Exoplanet Survey Satellite (TESS).

TOI 700 d orbits its star at a distance of 0.163 AU (24,400,000 km) from its host star with an orbital period of roughly 37.4 days, has a mass 1.72 times that of Earth, and has a radius of around 1.19 times that of Earth. It has been estimated that the planet receives about 86% the energy that the Earth receives from the Sun.

OVERALL COMPARATIVE ANALYSIS

According to the **size criteria**, the closest planetary **mass** objects by known radius or mass are shown in Table 10.1:

TABLE 10.1
Comparative Analysis of Exoplanets According to Size

Name	Earth masses ($M_{\oplus}$)	Earth radii ($R_{\oplus}$)	Note
Kepler-69c	0.98	1.7	Originally thought to be in the circumstellar habitable zone (CHZ), now thought to be too hot.
Kepler-9d	>1.5[24]	1.64	Extremely hot.
COROT-7b	<9	1.58	
Kepler-20f	<14.3[22]	1.03[22]	Slightly larger and likely more massive, far too hot to be Earth-like.
Tau Ceti b	2		Extremely hot. Not known to transit.
Kepler-186f		1.1[25]	Orbits in the habitable zone.
Earth	1	1	Orbits in the habitable zone.
Venus	0.815	0.949	Much hotter.
Kepler-20e	<3.08[21]	0.87[21]	Too hot to be Earth-like.
Proxima b	>1.27	>1.1	Closest exoplanet to Earth.
TRAPPIST-1 1c,1d,1e	1e–0.7	0.92	Close to earth 40 LY away. Rocky 6.1-day orbit time

This comparison indicates that size alone is a poor measure, particularly in terms of habitability. Temperature must also be considered as Venus and the planets of Alpha Centauri B (discovered in 2012), Kepler-20 (discovered in 2011), COROT-7 (discovered in 2009) and the three planets of Kepler-42 (all discovered in 2011) are very hot, and Mars, Ganymede and Titan are frigid worlds, resulting also in a wide variety of surface and atmospheric conditions. The masses of the solar system's moons are a tiny fraction of that of Earth, whereas the masses of extrasolar planets are very difficult to accurately measure. However, discoveries of Earth-sized terrestrial planets are important as they may indicate the probable frequency and distribution of Earth-like planets.

The goal of current searches is to find Earth-sized planets in the habitable zone of their planetary systems (also sometimes called the *Goldilocks zone*). Planets with oceans could include Earth-sized moons of giant planets, though it remains speculative whether such 'moons' really exist. The Kepler telescope might be sensitive enough to detect them. There is speculation that rocky planets hosting water may be commonplace throughout the Milky Way.

PROBABLE EARTH 2.0

TRAPPIST-1 PLANETS

TRAPPIST-1, as explained above, is located within the red circle in the constellation Aquarius (the Water Carrier). The star at the centre of the system was discovered in 1999 during the Two Micron All-Sky Survey (2MASS). It was entered in the subsequent catalogue with the designation "2MASS J23062928-0502285".

TRAPPIST-1 star and seven planets: The inner three worlds were discovered in 2016 by astronomers using the Transiting Planets and Planetesimals Small Telescope (TRAPPIST) at the European Southern Observatory in Chile. The outer four planets were spotted a year later, and it is possible that all seven worlds could potentially be habitable.

Now, new results have narrowed down the masses of the planets, confirming that they are all likely to be rocky, without the extended atmospheres that miniature versions of Uranus and Neptune would have. Furthermore, five of the planets have densities that suggest a significant amount of water, which is vital for life as we know it.

Astronomers led by Simon Grimm of the Centre for Space and Habitability at the University of Bern in Switzerland have produced the most accurate calculations of the planetary masses yet by taking advantage of a phenomenon known as Transit Timing Variations, or TTVs. The challenge of disentangling the data to calculate the planetary masses required new code written to handle 35 different parameters, five for each exoplanet. These are mass; orbital period; eccentricity; the argument of perihelion (the angle between its perihelion position and where the exoplanet's inclined orbit passes through the ecliptic plane); and the mean anomaly (an angle required to calculate an exoplanet's location on its elliptical orbit at any given time).

The code produced a range of different solutions, from which Grimm's team determined the configuration that best fits the observational data.

The most massive of the seven worlds is exoplanet "c", the second from the star, with a mass 1.156 times that of Earth. The least massive is exoplanet d, with less than a third of Earth's mass. The magnitude of the transits tells astronomers the radii of the exoplanets, and from their radius and mass, their densities can be calculated.

This is where things start to become interesting. Based on their densities, all seven worlds are predominantly rocky but contain up to 5% water. This is much more water than Earth's oceans (which amount to just 0.02% of Earth's mass). However, it remains to be seen whether the TRAPPIST-1 water is present on the surface of the exoplanet in vast, deep oceans; whether it is as vapour in a dense, steamy atmosphere; or whether it is spread around inside the exoplanet. Based on its temperature, exoplanet e would be the most like Earth according to Grimm.

This world has 77% of the mass of Earth, but is a little denser, indicating a large iron core and a thin atmosphere, possibly even thinner than Earth's.

TRAPPIST-1 "still represents the best opportunity we have for studying Earth-sized worlds outside of our own solar system", scientists confirm. The research is described in *Nature Astronomy* and in an upcoming paper in *Astronomy and Astrophysics*.

Proxima b

Proxima Centauri b is an exoplanet orbiting in the habitable zone of the red dwarf star Proxima Centauri, which is the closest star to the Sun and part of a triple star system. It is approximately 1.28 parsecs or 4.2 light years (4.0 × 10¹³ km) from Earth in the constellation Centaurus, making it and Proxima b the closest known exoplanets to the solar system (Figure 10.7).

Proxima Centauri b orbits the star at a distance of roughly 0.05 AU (7,500,000 km) with an orbital period of approximately 11.2 Earth days, and it has an estimated mass of at least 1.2 times that of Earth. It is subject to stellar wind pressures of more

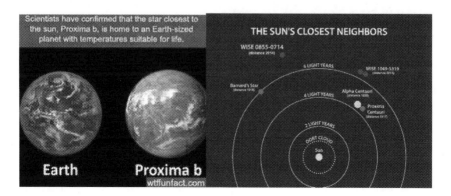

FIGURE 10.7   Proxima b compared with Earth.

than 2000 times those of Earth from the solar wind, and its habitability has not yet been definitively established.

The planet's discovery was announced in August 2016. It was found using the radial velocity method, where periodic Doppler shifts of the parent star's spectral lines suggest an orbiting object. From these readings, the parent star's radial velocity relative to the Earth is varying with an amplitude of about 1.4 m per second. According to Guillem Anglada-Escudé (Spanish Astronomer), the planet's proximity to Earth offers an opportunity for robotic space exploration with the star's hot project or at least "in the coming centuries".

Without its orbital inclination known, Proxima Centauri b's exact mass is unknown. If its orbit is nearly edge-on, it would have a mass of $1.173 \pm 0.086\ M_\oplus$ (Earth mass). Earth's mantle contains the equivalent amount of water as in the oceans etc. Statistically, there is a roughly 90% chance that its mass is less than $2.77\ M_\oplus$.

TOI 700 d

The time has come to analyse all aspects of living suitability. More than 4500 planets have been identified in the habitable zones of different stars, several light years away from our Sun. NASA's space exoplanet research programs consist of orbiting **space telescopes** studying the universe, along with ESA's telescopes.

Epilogue

NEED FOR GLOBAL SPACE VISION FOR SAVING EARTH

"...Our planet is a lonely speck in the great enveloping cosmic dark. In our obscurity, in all this vastness, there is no hint that help will come from elsewhere to save us from ourselves." -Carl Sagan

History proved that war is humanity's curse to prove the supremacy of nations. It may be based on ethnicity, territory, or religion; used to demonstrate a superpower position or survival of political leaders; or due to militancy. War heroes are worshipped, but all empires are short lived and have gone. Earth has transformed from a gladiator's world to more civilised, with democracy in most nations. Still, conflicts exist over border disputes, land/sea claims, and military domination in certain pockets of the world.

One should learn from Europe. The countries who once fought found a way to live together through the European Union. Less poverty and smaller population are also contributors, along with political will, an industrial economy, and technology adaptations for better living. But such industrially developed nations, along with America, have contributed the most to global warming and the ozone hole over Antarctica. Several decades of discussions in G7 and G20 did not make the Earth happy. Nearly three billion people in Asia, Africa, and South America are in poverty with limited infrastructure for health, education, employment, and housing in most regions. Rich countries who ruled them earlier and took away their prosperity did not help them, except to sell their arms and make their own industry work. It is the turn of the rich nations to pool resources and pump them into these countries to provide for the needs of those three billion people to have a reasonably good life. In the name of help, rich nations should not do business. Unfortunately, the UN has no teeth to force big nations to contribute substantially for the development of financially weaker nations, for a better life, eliminating poverty from the Earth.

The essence of the book is not only to impart knowledge on rocket science to students, but also to understand the challenge to protect our own Earth. All nations must come together with one vision for the long-term survival of humanity. Earth should not become barren like Mars. As Carl Sagan assessed, Earth is the only planet known to harbour life, and it is our duty to protect Earth for future generations. Therefore, humanity needs a great vision to forget all the conflicts, differences, superpower ambitions, military build-ups, and border issues, and move towards a common goal of peace and prosperity for all global citizens. Tamil Poet said 2000 years ago, YADUM OORE YAVARUM KELIR, meaning all nations are one; all people belong to one family.

The single vision of humanity must be to make Planet Earth liveable for generations.

For many generations, as long as the Sun shines, humanity must live on the Earth in peace, happiness, and prosperity with adequate drinking water, pollution-free air, organic pesticide-free agriculture, good education, employment, and healthcare for all.

The ozone hole over Antarctica must be plugged. Space technology must be for the development of people only. Space must be free from military intentions. Debris must be removed so that **Earth can smile at us.** Forty billion trees must be planted to convert 51 billion tons of CO_2 being produced every year to oxygen, throughout the world.

This vision for humanity to save Mother Earth for the life of future generations must be greater than any other vision so far envisioned by any nation. All nations must come together to pledge and act on this vision.

Glossary

TECHNICAL TERMS THAT FALL INTO THIS CATEGORY, AND THEREFORE APPEAR REPEATEDLY, ARE DEFINED HERE

Action: A force (push or pull) acting on an object.

Active controls: Devices on a rocket that move to control the rocket's direction in flight.

Antiballistic missile (ABM): A missile designed to intercept and destroy ballistic missiles or their warheads in flight. Antiballistic missiles are one form of missile defence system and, at this writing, the only form to be tested under real-world conditions.

Attitude control rockets: Small rockets that are used as active controls to change the attitude (direction) a rocket or spacecraft is facing in outer space.

Ballistic missile: A missile that travels in a long, parabolic trajectory, arcing upward under power and then falling back through the atmosphere to strike its intended target.

Black powder: A mixture of charcoal, sulphur, and saltpeter (potassium nitrate) similar, but not identical, to gunpowder. Black powder was the standard rocket propellant from the Middle Age to the early 20th century.

Booster: Technically, a self-contained rocket motor designed to be attached to a rocket or missile to give it extra power at liftoff. It is also used, casually, to refer to a launch vehicle, as in "PSLV Booster", "Atlas Booster," or "Soyuz Booster."

Canards: Small movable fins located towards the nose cone of a rocket.

Case: The body of a solid propellant rocket that holds the propellant.

Centre of mass (CM): The point in an object about which the object's mass is centred.

Centre of pressure (CP): The point in an object about which the object's surface area is centred.

CEP: Acronym for "circular error probable," a measure of a missile's accuracy. The CEP of a missile is the radius of a circle, centred on a given target, within which 50% of the missiles fired at that target will land.

Chamber: A cavity inside a rocket where propellants burn.

Combustion chamber: The part of a liquid-propellant rocket engine in which fuel and oxidiser are combined and burnt.

Cold launch: A launch technique used aboard ships and, in some land-based missile silos, that ejects the missile from its storage container with a charge of compressed gas. The missile's own motor ignites only when it is well clear of its container. Compare to "hot launch."

211

Cruise missile: A guided missile shaped like a small airplane. A typical cruise missile has some form of wings and tail and uses a jet engine as its primary propulsion system (though it may also have a booster rocket for takeoffs).

Drag: Friction forces in the atmosphere that "drag" on a rocket to slow its flight.

Escape velocity: The velocity an object must reach to escape the pull of Earth's gravity.

Extravehicular activity (EVA): Spacewalking.

Fins: Arrow-like wings at the lower end of a rocket that stabilise the rocket in flight.

Fuel: The chemical that combines with an oxidiser to burn and produce thrust. One of two principal components of a rocket's propellant (the other is oxidiser). Early liquid-propellant rockets used fuels such as gasoline, kerosene, and alcohol. Early solid fuel rockets of the 1940s used asphalt and similar compounds. "Fuel" is sometimes used, casually, as a synonym for "propellant," as in "liquid-fuel and solid-fuel rockets."

Gimballed nozzles: Tiltable rocket nozzles used for active controls.

Guidance system: The electrical and/or mechanical system that steers a missile while in flight. The type of guidance system used in a missile varies according to the desired mission, the size of the missile, and the technology available. Guidance system mechanisms fall into several categories. One type (heat-seeking, radar-seeking) locks the missile directly onto the target. A second (inertial) maintains the missile on a preprogrammed course by measuring and correcting its deviation from that course. A third (satellite, terrain-following) steers the missile by reference to an outside source.

Hot launch: A launch in which a missile is propelled out of its silo by its own exhaust gases. Compare to "cold launch."

ICBM: Acronym for "intercontinental ballistic missile": a missile with a range greater than 5500 km.

Igniter: A device that ignites a rocket's engine(s).

Injectors: Showerhead-like devices that spray fuel and oxidiser into the combustion chamber of a liquid propellant rocket.

Insulation: A coating that protects the case and nozzle of a rocket from intense heat.

IRBM: Acronym for "intermediate range ballistic missile": a missile with a range between 3000 and 5500 km.

Kiloton (KT): A unit used to measure the explosive power, or "yield," of nuclear weapons, equivalent to 1000 tons of the conventional explosive TNT. Compare to "megaton."

Launch vehicle: A rocket-powered vehicle, steered by a guidance system that is designed to carry a satellite or spacecraft from Earth's surface into space.

Liquid propellant: Rocket propellants in liquid form.

Mass: The amount of matter contained within an object.

Mass fraction (MF): The mass of propellants in a rocket divided by the rocket's total mass.

Megaton (MT): A unit used to measure the explosive power, or "yield," of nuclear weapons, equivalent to one million tons (1000 kilotons) of the conventional explosive TNT.

Microgravity: An environment that imparts to an object a net acceleration that is small compared with that produced by Earth at its surface.

MIRV: Acronym for "multiple, independently targeted reentry vehicle." A single missile carrying MIRVs can strike multiple targets. The acronym is also used as an adjective ("The treaty limits MIR-ICBMs") and as a verb ("If Russia also chooses to MIRV its SLBMs, the treaty will be in jeopardy").

Missile: A rocket-powered vehicle steered by a guidance system and used as a weapon. Compare to "launch vehicle" and "rocket".

Motion: Movement of an object in relation to its surroundings.

Movable fins: Rocket fins that can move to stabilise a rocket's flight.

Nose cone: The cone-shaped front end of a rocket.

Nozzle: A bell-shaped opening at the lower end of a rocket through which a stream of hot gases is directed.

Oxidiser: A chemical containing oxygen compounds that permits rocket fuel to burn both in the atmosphere and in the vacuum of space. One of the two principal components of rocket propellant (the other is fuel). Oxidiser supplies the oxygen necessary for the fuel to burn, which enables rockets to work in a vacuum. Jets and other "air-breathing" engines use the atmosphere as a source of oxygen.

Passive controls: Stationary devices, such as fixed rocket fins, that stabilise a rocket in flight.

Payload: The "discretionary" cargo (scientific instruments, satellites, spacecraft, etc.) that a rocket-powered vehicle carries or, more specifically, the weight of cargo it *can* carry. The payload of a missile – the total weight of guidance system and warhead that it can carry – is often called its "throw-weight." Payload is a function of thrust, and the payload of a given missile or launch vehicle varies according to the range or altitude desired.

Propellant: A mixture of fuel and oxidiser that burns to produce rocket thrust. An umbrella term for the fuel and oxidiser that power a rocket. Propellants may be solid (in which case the fuel and oxidiser are mixed before being placed in the rocket casing) or liquid (in which case they are carried separately and mixed in the combustion chamber just before ignition).

Pumps: Machinery that moves liquid fuel and oxidiser to the combustion chamber of a rocket.

Range: 1) The maximum horizontal distance that a given rocket or missile can travel; a function of thrust and payload. 2) An uninhabited area over which rockets, and missiles are fired for testing and research, such as ISRO's Sriharikota rocket launch station, DRDO's Integral Test Range, Chandipur-at-sea.

Reaction: A movement in the opposite direction from the imposition of an action.

Reentry vehicle: The separable nose section of a ballistic missile, designed to shield the warhead (and make it easier to steer) as it falls toward the target. See also "MIRV."

Regenerative cooling: Using the low temperature of a liquid fuel to cool a rocket nozzle.

Rest: The absence of movement of an object in relation to its surroundings.

Rocket: 1) A machine that is accelerated by accelerating a stream of particles (the working fluid) in the opposite direction. To date, nearly all rockets have used the hot gases produced by burning propellants as a working fluid. See also "rocket motor" and "thruster." 2) A rocket-powered device without a guidance system, designed to be used as a weapon. Compare to "missile."

Rocket motor: A self-contained rocket designed to be installed as a propulsion system for a vehicle such as an airplane, spacecraft, or missile.

SAM: Acronym for surface-to-air missile: a missile fired from the ground at enemy aircraft. There are also air-to-air missiles (AAMs), air-to-surface missiles (ASMs), and surface-to-surface missiles (SSMs), but SAM is the only one of the four acronyms to become a word in its own right. SAM is also used for safe-arming-mechanism in warhead.

SLBM: Acronym for submarine-launched ballistic missile: a missile designed to be carried aboard and launched from specially designed submarines.

Solid propellant: Rocket fuel and oxidiser in solid form.

Sounding rocket: A rocket-powered vehicle designed to carry small payloads such as cameras or scientific instruments into the upper layers of the atmosphere.

Spacecraft: A vehicle with guidance- and attitude-control systems, designed to operate in space. The category encompasses vehicles with and without human crews, vehicles capable of leaving Earth under their own power (like the space shuttle), and vehicles that must be lifted to orbit aboard launch vehicles (like PSLV, Apollo, Soyuz, or planetary probes).

Specific impulse: The amount of thrust generated by one kg of fuel in one second; a measure of the efficiency of a rocket motor.

Stages: Two or more rockets stacked on top of each other in order to reach higher altitudes or have a greater payload capacity.

Throat: The narrow opening of a rocket nozzle.

Thrust: The force that moves a rocket forward, usually expressed in newtons.

Thruster: A term sometimes applied to rockets that do not use the combustion of propellants to produce a working fluid. Small thrusters that use compressed gas as a working fluid have been used to control spacecraft and high-altitude research aircraft. Larger "ion thrusters" that use electricity to accelerate a stream of electrically charged atoms are now being developed as propulsion systems.

Unbalanced force: A force that is not countered by another force in the opposite direction.

Vernier rockets: Small rockets that use their thrust to help direct a larger rocket in flight.

Warhead: The destructive part of a missile's payload, consisting of the weapon itself (chemical explosives, nuclear explosives, toxic chemicals, or biological agents) and their associated fuses, triggers, and dispersal mechanisms.

Working fluid: The material, usually a gas that a rocket accelerates in order to produce thrust. Most rockets use the gases produced by the burning of their propellants as a working fluid.

References

REFERENCES (ALPHABETICAL ORDER)

1. Akash (missile) – Wikipedia. https://en.wikipedia.org/wiki/Akash_%28missile%29
2. Ancient Rishis. http://diehardindian.com/ancient-rishis/
3. Are robots going to replace humans? https://andreamangone.com/are-robots-going-to-replace-humans/
4. Asteroids – NASA Solar System Exploration. https://solarsystem.nasa.gov/asteroids-comets-and-meteors/asteroids/in-depth/?sa=X&ved=2ahUKEwigksHl08HxAhV0lGoFHTkyANQQ9QF6BAgHEAI
5. Asteroids. https://www.space.com/51-asteroids-formation-discovery-and-exploration.html
6. Astrobiology & Ocean Worlds. https://scienceandtechnology.jpl.nasa.gov/research/research-topics-list/planetary-sciences/astrobiology-ocean-worlds
7. Astronomy, Cosmology Science is Based on Vedas, Vedic.... https://haribhakt.com/cosmic-science-of-today-is-based-on-vedic-hindu-texts-written-thousands-of-years-ago/
8. AVTAR- RLV | India Defence. https://defenceprojectsindia.wordpress.com/2012/07/23/avtar-rlv/
9. Basic Principles of Satellite – KUET. http://www.kuet.ac.bd/webportal/ppmv2/uploads/1594994605Basic%20Of%20Satellite.pdf
10. Bhaskarachārya | Ancient Indian Science and Technology. https://ancientindianscience.net/the-great-ancient-indian-mathemetician-bhaskaracharya/
11. Big Bang https://www.coursehero.com/file/69721500/The-Big-Bangdocx/
12. Billion Year Plan: September 2011. https://billionyearplan.blogspot.com/2011/09/
13. Black Hole – Wikipedia. https://en.wikipedia.org/wiki/Black_hole
14. Black Holes. https://phys.libretexts.org/Bookshelves/Astronomy__Cosmology/Big_Ideas_in_Cosmology_(Coble_et_al.)/11%3A_Black_Holes
15. BrahMos: Supersonic Cruise Missile. https://pmso.in/brahmos-supersonic-cruise-missile/
16. Brief History of Space Exploration | The Aerospace.... https://aerospace.org/article/brief-history-space-exploration
17. CERN – Wikipedia. https://en.wikipedia.org/wiki/European_Organization_for_Nuclear_Research
18. Climate Action. https://www.helpxchange.org/climate-action
19. Climate Change https://www.agnb-vgnb.ca/content/dam/agnb-vgnb/pdf/Reports-Rapports/2017V1/Chap3e.pdf
20. Climate Change: Vital Signs of the Planet. https://climate.nasa.gov/causes/
21. Cryogenic Rocket Engine – Wikipedia. https://en.wikipedia.org/wiki/Cryogenic_rocket_engine
22. Dark Energy, Dark Matter | Science Mission Directorate. https://science.nasa.gov/astrophysics/focus-areas/what-is-dark-energy
23. Dark Energy. http://chandra.harvard.edu/xray_astro/dark_energy/
24. Dark Matter. https://starchild.gsfc.nasa.gov/docs/StarChild/universe_level2/darkmatter.html
25. Dark Matter?. https://www.space.com/20930-dark-matter.html
26. Design and Analysis on Scramjet Engine Inlet. http://www.ijsrp.org/research-paper-1301/ijsrp-p1335.pdf

27. Earth Observation Satellites – ISRO. https://www.isro.gov.in/spacecraft/earth-observation-satellites
28. Earthquake – Protection, Definition, Causes, Effects.... https://byjus.com/physics/protection-against-earthquake/
29. Escape Velocity – Wikipedia. https://en.wikipedia.org/wiki/Escape_velocity
30. Evolution – The Theory of Evolution by Natural Selection.... https://www.coursehero.com/file/49885604/evolutionpdf/
31. Extinction Level Event. https://www.crystalinks.com/ELE.html
32. Factoria Space Exploration Rocket Fuel. https://uploads.strikinglycdn.com/files/600faf6b-3cd2-4735-ace0-5a84eb79e1ad/45095197270.pdf
33. Farming Technology Use. https://eos.com/blog/top-5-newest-technologies-in-agriculture/
34. Future Communication Satellite. https://www.nextbigfuture.com/2019/03/future-communication-satellite.html
35. Future of Space Exploration – Wikipedia. https://en.wikipedia.org/wiki/Future_of_space_exploration
36. General relativity - Wikipedia. https://en.wikipedia.org/wiki/Introduction_to_general_relativity
37. Geocentric Orbit – Wikipedia. https://en.wikipedia.org/wiki/Earth_orbit
38. Great Indian Hindu Sages. https://www.hindujagruti.org/articles/31.html
39. Guidance System – NASA. https://www.grc.nasa.gov/WWW/k-12/rocket/guidance.html
40. Helium-3. https://mdcampbell.com/Helium-3version2.pdf
41. Hindu Cosmology. https://www.hinduscriptures.com/vedic-sciences/hindu-cosmology/27472/
42. History and Applications of Communication Satellites. https://www.edinformatics.com/inventions_inventors/communication_satellite.htm
43. History of Earth – Wikipedia. https://en.wikipedia.org/wiki/History_of_Earth
44. Hohmann Transfer Orbit – Wikipedia. https://en.wikipedia.org/wiki/Hohmann_transfer
45. How Climate Change and Global Warming Works. https://www.joboneforhumanity.org/global_warming
46. How Light Metals Help SpaceX Land Falcon 9 Rockets with.... https://www.lightmetalage.com/news/industry-news/aerospace/how-light-metals-help-spacex-land-falcon-9-rockets-with-astonishing-accuracy/
47. How Vera Rubin Discovered Dark Matter. https://astronomy.com/news/2016/10/vera-rubin
48. Hybrid-Propellant Rocket – Wikipedia. https://en.wikipedia.org/wiki/Hybrid-propellant_rocket
49. Hypersonic Airbreathing Propulsion – jhuapl.edu. https://www.jhuapl.edu/Content/techdigest/pdf/V26-N04/26-04-VanWie.pdf
50. Hypersonic Airbreathing Propulsion – jhuapl.edu. https://www.jhuapl.edu/Content/techdigest/pdf/V26-N04/26-04-VanWie.pdf
51. Ideal Rocket Equation. https://www.grc.nasa.gov/WWW/k-12/rocket/rktpow.html
52. International Space Station. https://artsandculture.google.com/entity/international-space-station/m03wky?hl=en
53. International Space Station. https://officerspulse.com/international-space-station-iss/
54. Introduction to General Relativity. https://amedleyofpotpourri.blogspot.com/2014/08/introduction-to-general-relativity.html
55. ISRO's Scramjet Engine Technology Demonstrator.... https://www.isro.gov.in/launchers/isro%E2%80%99s-scramjet-engine-technology-demonstrator-successfully-flight-tested

56. Kepler Telescope Discovers First Earth-Size Planet in Habitable Zone. https://www.nasa.gov/ames/kepler/nasas-kepler-discovers-first-earth-size-planet-in-the-habitable-zone-of-another-star

57. Kepler's Laws of Planetary Motion | Definition, Diagrams.... https://www.britannica.com/science/Keplers-laws-of-planetary-motion

58. Kepler-186f Is an Exoplanet Orbiting the Red Dwarf Kepler.... https://www.gettyimages.es/detail/fotograf%C3%ADa-de-noticias/kepler-186f-is-an-exoplanet-orbiting-the-red-fotograf%C3%ADa-de-noticias/973203020

59. Kepler-22b – Wikipedia. https://en.wikipedia.org/wiki/Kepler_22_B

60. Kepler-62f, a Small Habitable Zone World (Artist Concept). https://www.jpl.nasa.gov/images/kepler-62f-a-small-habitable-zone-world-artist-concept

61. Konstantin Tsiolkovsky and Activities in Outer Space. https://www.spacelegalissues.com/space-law-konstantin-tsiolkovsky/

62. Liquid Propellants – Engineering108.com. http://www.engineering108.com/Data/Engineering/aeronautical_engineering/Rocket_Propulsion_Elements/26429_07_engineering108.com.pdf

63. Mars 2020 Missions. https://cosmospnw.com/2020-mars-missions/

64. Medium Earth orbit. https://www.assignmentpoint.com/other/medium-earth-orbit-region-of-space.html

65. Milky Way – Wikipedia. https://en.wikipedia.org/wiki/Milky_Way

66. NASA Solar System.... https://solarsystem.nasa.gov/news/335/10-things-all-about-trappist-1/

67. NASA: 60 Years and Counting – The Future. https://www.nasa.gov/specials/60counting/future.html

68. NASA: Back to the Moon. https://www.nasa.gov/specials/apollo50th/back.html

69. NASA's Kepler Discovers Its Smallest 'Habitable Zone.... https://www.nasa.gov/mission_pages/kepler/news/kepler-62-kepler-69.html

70. New Earth-like Planets Found | Carnegie Institution for.... https://carnegiescience.edu/news/new-earth-planets-found

71. Newton In Space – NASA. https://er.jsc.nasa.gov/seh/Newton_In_Space.pdf

72. Newton's Law of Universal Gravitation. https://web2.ph.utexas.edu/~tsoi/303K_files/ch9.pdf

73. Newton's Law of Universal Gravitation – Wikipedia. https://en.wikipedia.org/wiki/Newton%27s_law_of_gravitation

74. Orbit. https://www.nasa.gov/audience/forstudents/5-8/features/orbit_feature_5-8.html

75. Origins of the Universe. https://www.berkeley.edu/news/media/releases/2007/03/16_hawking_text.shtml

76. Planets in Our Solar System: Names & Features. https://healthywaymag.com/science/planets-solar-system

77. Polar Orbit. https://nasa.fandom.com/wiki/Polar_orbit

78. Practical Rocketry – NASA. https://www.grc.nasa.gov/WWW/k-12/rocket/TRCRocket/practical_rocketry.html

79. Precision Agriculture: How to Improve Farming https://eos.com/blog/precision-agriculture-from-concept-to-practice/

80. Quasar — Wikipedia. https://en.wikipedia.org/wiki/Quazar

81. Ramjet – Wikipedia. https://en.wikipedia.org/wiki/Ramjet

82. Rocket History – web.eng.fiu.edu. http://web.eng.fiu.edu/allstar/Rock_Hist1.html

83. Rocket Principles – NASA. https://www.grc.nasa.gov/WWW/k-12/rocket/TRCRocket/rocket_principles.html

84. Rocket Principles A – NASA. https://er.jsc.nasa.gov/seh/04_Rocket_Principles.pdf

85. Rockets. https://estesrockets.com/wp-content/uploads/Educator/LessonPlans/ERL_SS_3-5_3.pdf

86. Satellite – Wikipedia. https://en.wikipedia.org/wiki/Satellites
87. Satellite Communications. https://www.amitel.com/satellite-communications/
88. Satellite Navigation – Wikipedia. https://en.wikipedia.org/wiki/Satellite_navigation
89. Satellite. https://www.nasa.gov/audience/forstudents/5-8/features/what-is-a-satellite-58_prt.htm
90. Scope of Mining on the Moon. http://dspace.nitrkl.ac.in/dspace/bitstream/2080/1701/1/pAPER4-Mining%20The%20Moon-raipur-nov2011-MINTECH11.pdf
91. Scramjet – Wikipedia. https://en.wikipedia.org/wiki/Scramjet
92. Space debris – Wikipedia. https://en.wikipedia.org/wiki/Space_debris
93. Space Exploration and Mars colonization. https://www.gtmars.com/space_exp.html
94. Space Junk Getting Worse. https://slashdot.org/story/10/02/24/1634224/space-junk-getting-worse
95. Space Shuttle – Wikipedia. https://en.wikipedia.org/wiki/Space_Shuttle
96. Space Waste. http://dictionary.sensagent.com/SPACE%20WASTE/en-en/
97. SpaceX Wikipedia. https://en.wikipedia.org/wiki/Merlin_(rocket_engine_family)
98. SpaceX Raptor – Wikipedia. https://en.wikipedia.org/wiki/Raptor_2
99. Specific Impulse – Wikipedia. https://en.wikipedia.org/wiki/Specific_impulse
100. Sputnik. https://www.history.nasa.gov/sputnik.html
101. Stellar Evolution – Wikipedia. https://en.wikipedia.org/wiki/Life_cycle_of_a_star
102. Stephen Hawking Travels Back in Time. https://www.laphamsquarterly.org/time/stephen-hawking-travels-back-time
103. String Theory – Wikipedia. https://en.wikipedia.org/wiki/String_Theory
104. Sun V Wikipedia. https://en.wikipedia.org/wiki/Solar_diameter
105. Sustainability for Reaching the Bottom. https://indiacsr.in/sustainability-for-reaching-the-bottom-of-the-pyramid-by-dr-apj-abdulkalam/
106. Sustainable Development. https://www.ugb.ro/etc/etc2014no1/03_Bontas_D..pdf
107. The Causes of Climate Change. https://awaken.com/2020/08/the-causes-of-climate-change/
108. The Dangers of Space Junk – Space Waste Solutions. https://solutionsforspacewaste.com/the-dangers-of-space-junk/
109. The Four Fundamental Forces of Nature | Space. https://www.space.com/four-fundamental-forces.html
110. The Future of Humanity: Our Destiny in the Universe. https://www.barnesandnoble.com/w/the-future-of-humanity-michio-kaku/1126840812
111. The History of Early Fireworks, Rockets and Weapons of War. https://www.thoughtco.com/early-fireworks-and-fire-arrows-4070603.
112. The History of Space Exploration | National Geographic Society. https://www.nationalgeographic.org/article/history-space-exploration/12th-grade/
113. The Milky Way. https://wikithat.com/wiki/a1bfe483-f763-4fe0-a06a-ae7f1abba6ec/The_Milky_Way_---/Milky_Way_(Our_Galaxy)
114. The Origin of the Universe. https://www.scholastic.com/teachers/articles/teaching-content/origin-universe/
115. The Weird Mystery of Dark Energy. https://astronomy.com/magazine/2018/07/the-weird-mystery-of-dark-energy
116. Tiangong Space Station – Wikipedia. https://en.wikipedia.org/wiki/Chinese_large_modular_space_station
117. Trees Improve Our Air Quality. http://urbanforestrynetwork.org/benefits/air%20quality.htm
118. Visualizing All of Earth's Satellites: Who Owns Our Orbit?. https://www.visualcapitalist.com/visualizing-all-of-earths-satellites/

BOOKS

1. Raj, Gopal. *Reach for the Stars – The Evolution of India's Rocket Programme.* Viking, 2000.
2. Pillai, A. Sivathanu. 'Future ISRO Launchers – A study' (Report No. ILV/S: TN: 32: 85/S).
3. Kalam, A.P.J. Abdul and Pillai, A. Sivathanu. 'Performance of Cost Effectiveness of ISRO Launchers' (Report No. ILV/S: RN:03:83).
4. Pillai, A. Sivathanu. 102 Launch Azimuth for Geosynchronous Mission from SHAR (Report No. ILV/S:TN:83).
5. Pillai, A. Sivathanu. PSLV-A Launch for Indian Meteorological Satellite.
6. Pillai, A. Sivathanu. 'HQ Analysis of PSLV Configuration Options' (Report No. PSLV: MRLCHR: 06: 91).
7. Gruntman, Mike. Blazing the Trail – The Early History of Spacecraft and Rocketry.
8. Roddam, Narasimha. Rockets in Mysore, and Britain, 1750–1850 AD, Project Document DU 8503, National Aeronautical Laboratory, Indian Institute of Science, Bangalore.
9. Kalam, A.P.J. Abdul. The History of Indian Rocketry, Hazrat Tipu Sultan Shaheed Memorial Lecture, November 30, 1991.
10. www.meddev.nic.in/science/policy.htm
11. Mistry, Dinshaw. *"Containing Missile Proliferation," Strategic Technology, Security, Regimes, and International Cooperation in Arms Control.* University of Washington Press.2003.
12. Mallik, Amitav. Technology and Security in the 21st Century, SIPRI Research Report No. 20.
13. McGraw-Hill Encyclopaedia of Science and Technology, Seventh Edition, Vol. 18, p. 151, 1992.
14. Mar, B.W., W.T. Newell, and B.O. Saxberg. Managing High Technology. An Interdisciplinary Perspective. Elsevier Science BV. 1985.
15. Subba Rao, A. *Management of Technology Change.* Global Business Press, 1994.
16. McCutcheon, Scott and Bobbi. *Space And Astronomy – The People behind Science.* Chelsea House. 2007.
17. Kalam, A.P.J. Abdul and Pillai, A. Sivathanu. *Thoughts for Change – We Can Do It.* Pentagon Press New Delhi, 2013.
18. Pillai, A. Sivathanu. *Success Mantra of Brahmos –The Path Unexplored.* Pentagon Press, 2014.
19. Planet Aerospace Veterans. *Quintessence of Nano-satellite Technology.* Notion Press, 2020.
20. Sutton, George P. and Oscar Biblarz. *Rocket Propulsion Elements.* John Wiley & Sons, 2001.
21. Bate, Roger, Donald D. Mueller, and Jerry E. White. *Fundamentals of Astrodynamics.* Dover Publications.1971.
22. Thomson, William. *Tyrrell Introduction to Space DynamIcs.* Dover, 1986.
23. Campbell, Bruce A. and Samuel Walter McCandless. *Introduction to Space Sciences and Spacecraft Applications.* Gulf Professional Publishing, 1996.
24. Turner, Martin J.L. *Rocket, and Spacecraft Propulsion: Principles, Practice and New Developments.* Springer Science & Business Media, 2008.
25. Brown, Charles D. *Spacecraft Mission Design.* AIAA, 1998.
26. Logsdon, T. *Orbital Mechanics: Theory and Applications.* John Wiley & Sons, 1997.
27. Larson, Wiley J., Gary N. Henry, and Ronald W. Humble, eds. *Space Propulsion Analysis and Design.* McGraw-Hill, 1995.

28. Larson, Wiley J. and James Richard Wertz. *Space Mission Analysis and Design.* Microcosm, 1992.
29. Taylor, Travis S. *Introduction to Rocket Science and Engineering.* CRC Press, 2009.
30. Sellers, Jerry Jon, et al. *Understanding Space: An Introduction to Astronautics.* Primis, 2000.
31. DRDO. *Integrated Guided Missile Development Programme.* DECIDOC, 2008.
32. Gupta, S.C. *Growing Rocket Systems and the Team.* Prism Books, 2006.
33. Nagappa, Rajaram. *Evolution of Solid Propellant Rockets in India.* DECIDOC, DRDO, 2014.
34. Rao, P.V. Manoranjan and P. Radhakrishnan. *A Brief History of Rocketry in ISRO.* Universities Press, 2012.
35. Neufeld, Michael j. *Von Braun –Dreamer of Space & Engineer of War.* Vintage Books – A Division of Random House, 2007.
36. Department of Air Force, USA. Soviet Aerospace Handbook. Department of Air Force.1978.
37. Hawking, Stephen and Leonard Mlodinow. *The Grand Design.* Bantam Books, 2011.
38. Hawking, Stephen. *The Universe in Nutshell.* Transworld publishers, 1988.
39. Hawking, Stephen and Roger Penrose. *The Nature of Space and Time.* Princeton University Press, 1995.
40. Sagan, Carl. COSMOS . Abacus, 1995.
41. McCutcheon, Scott and Bobbi. *Space & Astronomy –The People Behind the Science.* Viva Books, 2007.
42. Natarajan, Priyamvada. *Mapping the Heavens.* Yale University Press, 2016.
43. Christie, Latha. *Beyond the Boundaries of Science –Exploring the Cosmic Story.* KB Publishing, 2019.
44. Kaku, Michio. *The Future of Humanity.* Penguin Random House, 2018.
45. Kaku, Michio. *Visions.* Anchor Books, 1997.
46. Kaku, Michio. *Physics of the Future.* Penguin Books, 2011.
47. Regis, E.D.. The Biology of Doom: The History of America's Secret Germ Warfare Project. Henry Holt & Company.
48. Goswami, Amit. *God Is nNt Dead-Quantum Physics on Our Origins.* Jaico Publishing House, 2009.
49. Cox, Brian and Andrew Cohen. *Human Universe.* William Collins, 2014.
50. Penrose, Roger. *Cycles of Time –An Extraordinary New View of the Universe.* The Bodley Head, 2010.
51. Chun, Clayton K.S. *Thunder over the Horizon –From V2 Rockets to Ballistic Missile.* Pentagon Press, 2009.
52. Barrow, John D. *The Book of Universes.* The Bodley Head, 2011.
53. Dartnell, Lewis. *Origins –How the Earth Shaped Human History.* Penguin Random House, 2019.
54. Cham, Jorge and Daniel Whiteson. *We Have No Idea –A Guide to the Unknown Universe.* John Murray, 2017.
55. Einstein, Albert. *Relativity: The Special and General Theory.* Bonanza Books, 1961.
56. ISRO/SC/CIT/8/16 Report of PSLV Production.
57. Bakey, Ivan. *Advanced Space System Concepts as Technologies.* Aerospace Press, 2002.
58. Dave & others, *Results-Based Leadership.* HBS Press, 1999.
59. Angelo, Joseph A. *J R Rockets –Frontier in Space.* Facts on File, 2006.
60. van Riper, A. Bowdoin. *Rockets and Missiles –A Life Story of a Technology.* Greenwood Press, 2004.
61. Snedden, Robert. Rockets and Spacecraft. Twentieth Century Innovations. Wayland. 1991.

62. NASA – Rockets Educator Guide – 20.
63. Education Working Group. *NASA, Rocketry Basics*. Johnson Space Center.
64. Educational Guide. *Important Adventures in Rocket Science*. NASA, 2008.
65. III.4.2.1 Rockets and Launch Vehicles, Education Series NASA, Federal Aviation Administration. 2000.
66. van Pelt, Michel. *Rocketing into the Future –The History of Technology of Rocket Planes*. Springer, 2012.
67. Isakowitz, Steven J. *International Reference Guide to Space Launch Systems*. AIAA, 1999.
68. Haley, Andrew G. *Rocketry and Space Exploration*. Divan Nostrand Company, 1959.
69. D'Souza, Marsha R , Otalvaro, Diana M . Deep, Arjun Singh , Harvesting Helium-3 From the Moon – Project Report, Worcester Polytechnic Institute, Project Number: IQP-NKK-HEL3-C06-C06, 2006.
70. Natalie Lovegren, Chemistry on the Moon: The Quest for Helium-3, Fusion and Physical Chemistry – 21st Century Science and Technology.2014.
71. Space Debris and Challenges to Safety of Space Activity, Federal Space Agency of Russia, 2009.
72. UN COPUOS reports on Space Debris, UN. 2020.
73. UN COPUOS Report on Sustainable Development, European Space Policy Institute, 2016.
74. Gates, Bill. *How to Avoid a Climate Disaster*. Penguin Random House, 2021.
75. SPS – Space Propulsion Systems – Developing Environmentally Clean Technologies, 2007.
76. Number of Inputs from Internet, Wikipedia, Dr Kalam's lectures, website through Dr. V Ponraj – World Space Vision, 2007.
77. Pillai, A. Sivathanu. Space Enterprise – Core Competencies and Low-Cost Access, World Space-Biz 2010, Vikram Sarabhai and Satish Dhawan Memorial Lectures, Energy, Environment and Sustainable Development Lectures.
78. Srivathsa, BK , Narasimhan, MA . Science and Technology in India through the Ages, Academy of Sanskrit Research, 2003.
79. Kalam, A.P.J. Abdul and S.P. Singh. *Target 3 Billion –Innovative Solutions towards Sustainable Development*. Penguin Books, 2011.
80. Pillai, A. Sivathanu. *40 Years with Abdul Kalam –Untold Stories*. Pentagon Press, 2020.
81. Fleeman, Eugene L. Tactical Missile Design AIAA Education series, AIAA Education Series. 2001.

Index

6 DOF flight trajectory, 149

Acharya Bharadwaj, 7, 8
Acharya Kanad, 7
Acharya Kapil, 7
airbreathing propulsion systems, 127
aluminam-lithium casing, 105
Aryabhata, 6, 7
asteroids, 25, 35, 39, 40, 137, 167, 178, 179,
        180, 185
asteroids collision avoidance, 179
autonomous systems, 172, 173

Bhaskaracharya, 7
Big Bang Theory, 5, 14, 16
biosatellites, 143
Black hole, 9, 10, 12, 13, 21, 22, 23, 24,
        31, 43
Blue Dot-The Earth, 34
Blue Origin, 60, 143
BrahMos, xv, xix, 96, 97, 98
braneworld, 20
burn rate, 85, 86, 106
burst of supernova, 26

Carl Sagan, 2, 8, 39, 199
casing for liquid propellant stages, 105
centre of gravity, 69
centre of pressure, 69, 70
CERN, 9, 10, 14, 15, 16, 20
CFD, 99, 100, 115, 116, 117, 123
checkout system, 118, 119
Chinese fire arrows, 42
chlorofluorocarbons (CFCs), 186
clean energy generation, 197
control system, 70, 71
cosmology, 2, 3, 5, 7, 8, 10, 13, 26, 30
crewed spacecraft, 144
cryogenic rocket engines, 90, 91
Cygnus X-1, 24
cylindrical Hall thrusters, 158

dark energy, 4, 5, 25, 26, 27, 28, 150, 291
dark matter, 4, 5, 24, 25, 28, 29, 30,
        38, 150
digital platform, 195
DRDO, xv, xix, xxi, 97, 99, 168

early rocket designs, 44
Earth observation satellites, 56, 130, 139

Edwin Hubble, 9, 16, 31
Einstein Cross, 13
Einstein tensor, 10
Einstein's equations, 10
electric propulsion, 150, 151, 152, 153, 154, 156,
        157, 159, 160, 161
electromagnetic force, 18, 19, 34
electromagnetic/plasma engine, 153
electromagnetism, 18, 20
electrostatic thrusters, 153
electrothermal engine, 151
engine thrust control, 106
ESA, xxi, 9, 55, 90, 150, 178, 179, 182, 183,
        184, 208
exoplanets, 1, 40, 199, 200, 202, 203, 205, 207
external discharge Hall thruster, 158

fibre-optic gyroscopes (FOG), 112
fishing, 195
flex nozzle, 108, 109
Fred Hoyle, 14
fundamental physical forces, 17
future of Earth, 39

galaxy rotation curve, 29
general theory of relativity, 9, 12, 23
geo centric phase, 120
geo orbit, 124
geo stationary orbit, 133
geographic information systems, 194
geo-spatial pyramid, 197
global warming, 179, 184, 185, 187, 188,
        189, 190
God particle, 14, 16
grain shape, 86
gravitational lensing, 10, 12, 13
gravitational waves, 9, 10, 12, 13
gravity, 7, 9, 10, 12, 18, 19, 20, 21, 22, 24, 26, 29,
        30, 35
Great Indian Saints, 6

Hall effect thruster, 155
helio centric phase, 120
Hermann Oberth, 49
Higgs boson, 16
Hindu Puranas, 5
Hohmann Transfer Orbit, 120
Hubble Space Telescope, 55
Human Mission to Mars, 122
Human Space Missions, 146

hyperplane, 168, 169, 170, 171, 176
hypersonic transportation, 168

ideal rocket, 74, 76, 77, 78
importance of satellites, 130
inertial guidance system, 111, 112
Internet of Things, 194
ion thruster, 151, 154, 155, 157, 159, 160,
        161, 206
ISRO, xix, xx, xxi, 1, 57, 58, 78, 86, 88, 93, 99,
        120, 137, 139, 140, 168, 191

Kepler's Laws, 32
Kepler-186f, 199, 200
Kepler-22b, 202
Kepler-62e, 201, 202
Kepler-62f, 200
Kepler-69c, 202
killer satellites, 144
Konstantin Tsiolkovsky, xviii, xix, 2, 41
krypton, 155, 157

Large Hadron Collider, 15
laser propulsion, 149, 167
launch vehicle design methodology, 115
law of universal gravitation, 10
liquid propellant rockets, 49
liquid ramjet, 96, 97, 98
Low Earth Orbit (LEO), 165, 167, 169, 171
low-cost access to space, 170
lunar industry, 174

magneto plasma rocket propulsion, 160
man on moon, 54
Martian phase, 121
mass fraction, 101, 102
materials, 116, 117
Medium Earth Orbit (MEO), 133, 137, 140
MHD engine, 153
Milky Way, 31
mission planning, 137
Mother Earth, 210
multistage rockets, 45, 78
multi-universe, 30
MUONS, 20

nanotechnology, 159
NASA, xix, xxi, 4, 9, 40, 53, 54, 55, 58, 59, 60,
        61, 101, 123, 142, 150, 151, 155, 156,
        161, 173, 179, 180, 181, 182, 183, 200,
        202, 202, 203, 204, 208
navigation, 63, 98, 110, 111, 112
neutrinos, 16, 30
Newton's Laws of Motion, 65

Newton's First Law, 65
Newton's Second Law, 66
Newton's Third Law, 68
nuclear thermal rocket (NTR), 162, 163, 164, 165

orbiting satellite, 133, 136
ozone depletion, 187, 188

polar orbit, 133
Potable Water GRID, 197
precision farming, 192, 193, 194
pressure-fed system, 88, 89
Proxima Centauri b, 207, 208
PSLV, xix, xx, 57, 88, 103, 104, 109, 120

radioisotope thermoelectric generators, 174
ramjet, 94, 95, 96, 97
reconnaissance satellites, 133, 144
recovery satellites, 144
redshift, 16, 17, 26
Resitojet, 152
Rig Veda, 3, 5
Ring Laser Gyroscopes (RLG), 112
Robert H. Goddard, 47
robotic spacecraft development, 172, 174
rocket equation, 45, 46, 70, 74, 76, 77
rocket subsystems, 81
Ruscosmos, xxi, 55

satellite applications, 138
satellite attitude control, 109
satellite orbits, 132, 133
satellites in orbit, 137, 141
Saturn V, 92
scramjet, 94, 98, 99
semi-cryogenic rocket engines, 92
SHAR, 116, 124
SINS (Stablised platform INS), 111
small satellites, 141, 142, 143
solar electric propulsion (SEP), 165
solar sail, 149, 167, 168
solid propellant rocket, 78, 83, 86, 106
space odyssey, 51
space shuttle, 55
space stations, 47, 51, 54, 55, 183
space weather, 184
space-based solar power satellites, 144
space-time, 9, 10, 13, 18
SpaceX, 142, 143, 157
specific impulse, 90, 96, 97, 100, 109
Sputnik-1, 129
stability and control systems, 107
Starlink, 142
Star ship, 143

static testing, 86
stellar evolution, 21
Stephen Hawking, 9, 12, 23, 24
stereo vision for collision avoidance, 173
Strap Down Inertial Navigation System
        (SDINS), 112
string theory, 20, 21
strong force, 18, 19
supermassive Black Holes, 24

tachyons, 149
telescopes, 167, 199, 203, 204, 208
tether satellites, 144
the strong nuclear force, 19
the weak nuclear force, 19
threats from outer space, 178
Tiangong Space Station, 56
TOI 700 d, 205, 208
total impulse, 73
trajectory design, 118, 120, 122

TRAPPIST-1, 203, 204, 272
turbojet, 94, 98, 99

unifying nature, 19

Van Allen Radiation Belts, 124
Vikram Sarabhai, xv, xix, xx
Volatiles Investigating Polar Exploration Rover
        (VIPER), 58

water vapour, 185, 186
wind tunnel testing, 117, 123
World Food Program (WFP), 177
World launch sites, 124
World rockets, 61

xenon, 150, 151, 153, 154, 155, 156, 157, 160,
        161, 165

Yuri Gagarin, 53

Printed in the United States
by Baker & Taylor Publisher Services